# Mountain Nature

# Mountain Nature

## A SEASONAL NATURAL HISTORY

## *of the* SOUTHERN APPALACHIANS

JENNIFER FRICK-RUPPERT

The University of North Carolina Press    Chapel Hill

Manufactured in the United States of America

Set in Legacy and Stone Sans

Publication of this book was supported in part by generous gifts
from Frank and Doris Guest, Linda Martinson, Augustus and
Margaret Napier, Aleen Steinberg, and Kenneth and Harriet Walls.

The paper in this book meets the guidelines for permanence and
durability of the Committee on Production Guidelines for Book
Longevity of the Council on Library Resources.

The University of North Carolina Press has been a member of the
Green Press Initiative since 2003.

Frick-Ruppert, Jennifer.

  Mountain nature : a seasonal natural history of the Southern
Appalachians / Jennifer Frick-Ruppert.

      p. cm.

  Includes bibliographical references and index.

  ISBN 978-0-8078-3386-5 (cloth : alk. paper)

  ISBN 978-0-8078-7116-4 (pbk. : alk. paper)

  1. Natural history—Appalachian Region, Southern. 2. Forest
ecology—Appalachian Region, Southern. 3. Mountain ecology—
Appalachian Region, Southern. 4. Animal ecology—Appalachian
Region, Southern. I. Title.

  QH104.5.A6F75 2010

  508.75—dc22                                    2009039275

cloth   14 13 12 11 10   5 4 3 2 1
paper   14 13 12 11 10   5 4 3 2 1

*for Fritz*

# Contents

ACKNOWLEDGMENTS  xiii

INTRODUCTION: THE DIVERSE SOUTHERN APPALACHIANS  1

1  THE NATURE OF CYCLES  9

Daily Cycles and Biological Clocks  10
The Daily Cycle of Sleep in Animals  11
Daily Cycle of Sleep in Plants?  12
Communication at Night  12
Seasonal Cycles and Biological Calendars  15
Day Length as the Cue for Seasonal Calendars  17
Tree Rings: Evidence of Seasonal Cycles in Plants  19
Migration and Hibernation: Seasonal Cycles in Animals  20
Seasonal Cycles of Aquatic Animals  21
Appalachian Trout and Other Fish  23
Trout Food: Aquatic Insects  25
Longer-term, Multi-year Cycles  26
Tree Reproduction and Long-term Masting Cycles  27
Masting in Cicadas?  28
Population Peaks in Voles as Long-term Cycles  29
Geological Features as Extremely Long-term Cycles  31
The Nature of Cycles  33

2  CYCLES OF SPRING: MARCH, APRIL, MAY  35

Pollination and Flower Form  35
Pollination in Serviceberry and Silverbell  36
Wind Pollination in Maples, Oaks, and Grasses  37
Insect Pollination in Tulip Trees, Magnolias, and Flame Azaleas  40
Of Peas and Pollinators: Locusts and Other Legumes  41
Ephemeral Wildflowers: Ramps and Trout Lilies  44
Other Wildflowers: More Lilies, Trilliums, and Jack-in-the-Pulpits  47

"Leaves in Three, Let It Be!"  48

Oconee Bells: A Charismatic Appalachian Wildflower  49

Bloodroot: A Native Poppy  51

Plants and Ants: Beneficial Relationships  51

The Orchids: Wildflowers that Trick Insects and Parasitize Fungi  54

Morels: A Springtime Delicacy  56

Squawroot and Other Parasitic Plants  57

Honeybee Swarms: Colony Reproduction  59

Insects and Migratory Birds  61

Warbler Diversity and Decline  62

Caterpillars and Cuckoos  63

Blue Birds and Bluebirds  65

Migrations: The Return Route of Hummingbirds  66

Spring Peepers  68

Climatic Conditions in Spring  69

3 CYCLES OF SUMMER: JUNE, JULY, AUGUST  71

Flowers in Myriad Forms: Jewelweed and Dodder  71

Summer's Divine Robes: Cardinal Flower  74

Ginseng and Yellowroot: Uncommon and Common Medicinal Plants  75

Insectivorous Plants: Sundews and Pitchers  76

The Heaths: Mountain Laurel, Rhododendron, Sourwood, and Pinesap  78

Interdependence of Plants and Pollinators: Yucca  81

What Good Is a Mosquito?  82

Seeds and Fruits: Doll's Eyes and Hearts-a-Bustin'  83

Rain, Fungi, and Connections  84

Dangerous Social Insects and Their Mimics  87

"Unsocial" Wasps?  91

Poisonous Butterflies and Their Mimics  93

Moth Communication and Camouflage  96

Luna Moths: Endangered or Not?  98

Katydids and Crickets: Summer Songsters Use Sound to Communicate  99

Damsels and Dragons  101

Crab Spiders: Camouflage by the Predators  103

Flashing Fireflies of Summer Evenings  103

Railroad Worms and Millipedes: Predators and Prey  106

Grouse Threat Display: Prey or Predator?  108

Southern Appalachians: Greatest Salamander Diversity in the World  109

Lizards and Snakes: Close Cousins  111

Snake Sense  114

The Centenarian Turtles  117

Climatic Conditions in Summer  119

4 CYCLES OF FALL: SEPTEMBER, OCTOBER, NOVEMBER  121

Bioluminescent Mushrooms  122

Leaf Color Change  123

Fall Flowers: Gentians and Orchids  126

Witch Hazel's Bewitching Flowers  126

Dispersal of Offspring in Plants  128

Sweet Fleshy Fruits  130

Fatty Fruits? Spicebush and Dogwood  131

Other Fleshy Fruits: Sumac  133

Migrant and Resident Birds as Fruit Dispersers  134

Dry Fruits: Oaks, Hickories, and Chestnuts  135

Squirrels as Seed Dispersers  137

Bird Migration: Kinglets and Nightjars  138

Hawks: Migrants and Residents  140

Resident Screech and Barred Owls  143

Bats  144

Bats and Gargoyles  147

Migrating Monarchs and the Foods that Support Them  148

Ladybugs: Friends or Foes?  152

The Rise and Fall of Insect Populations: Woolly Bear Caterpillars  154

Woolly Alder Aphids  155

Spiders and Their Insect Prey  157

Climatic Conditions in Fall  161

5 CYCLES OF WINTER: DECEMBER, JANUARY, FEBRUARY  163

Freeze/Thaw Cycles and the Magic of Ice  163

Deciduous versus Evergreen Trees  164

Appalachian Conifers  167

The "Perfect Storm" of Acids, Adelgids, and Global Warming  168

American Holly: A Broad-Leaved Evergreen Tree  170

Rhododendrons as Evergreen Thermometers  171

Evergreen Herbaceous Plants: Cycles Reversed  172

Primitive Plants and Their Reproductive Cycles  173

Lichens and Jelly Fungi  176

Springtails: Enigmatic Winter Animals  178

Hibernation and Denning Cycles in Groundhogs and Bears 179

Activity Cycles of Small Mammals: Shrews, Moles, Mice,
   and Flying Squirrels 182

Common Nocturnal Omnivores: Raccoons, Skunks, and Opossums 185

Rarely Seen Large Carnivores: Foxes, Coyotes, Bobcats,
   Mountain Lions, and Otters 189

Eyeshine of Nocturnal Animals 192

Rabies Epidemics 192

Cycles of Bird Irruptions 193

Birdsong and the Cycles of Territoriality versus Foraging Flocks 195

Reproductive Cycles that Begin in Winter: Wood Frogs and
   Great Horned Owls 201

Climatic Conditions in Winter 204

## SIDEBARS

Effects of Global Warming 22

Fishing in Trout Rivers 27

Ramps Harvests 45

Fire Ants and Global Warming 53

Please Do Not Dig Orchids 55

Photovoltaic Cells: Humanity's Photosynthesis? 58

Caterpillar Circular Logic 64

Hummingbirds and Feeders 67

Poison Ivy Potion 72

Threats to Pitcherplants 78

Wasps as Friends, Not Foes 89

Treatment for Squash Borers 90

Silk Shirts from Worms! 99

Cool Night Lights 105

Fewer Fireflies? 106

Color Perception in Humans 125

Hawk Migration Sites 142

Bat Conservation 147

Monarch Migration Locations 151

Human Effects on Monarchs 152

Woolly Worms and Weather Predictions 155

Invasive Species 170

Old Coal and Fossil Fuels 175

Lichens as Indicators of Air Quality 178

Groundhog Day 180

Help Black Bears and People 181

"Bird" Feeders? 185

APPENDIX: Federal Public Lands in the Southern Appalachians 205

REFERENCES 207

INDEX 215

# Acknowledgments

Thanks to Dr. E. C. Bernard of the University of Tennessee for identifying the springtails, to Dr. R. M. Shelley of the North Carolina State Museum of Natural Sciences for identifying the millipedes, and to Dr. R. S. Fox of Lander University for identifying the cladocerans. I used A. S. Weakley's *Flora of the Carolinas, Virginia, Georgia, Northern Florida and Surrounding Areas*, available through the website of the University of North Carolina's herbarium, as the reference for all scientific names of plants. It is a valuable resource. The Western North Carolina Nature Center allowed me to photograph their animals.

For several years, I wrote a column called Appalachian Almanac for my local newspaper, *The Transylvania Times*, and many of these columns were revised and incorporated into this manuscript. Much of the manuscript was prepared during a sabbatical that was provided by Brevard College and funded in part by a grant from the Appalachian College Association.

Individuals who read the entire manuscript and offered suggestions, both scientific and editorial, included an anonymous reviewer, Dr. T. P. Spira of Clemson University, and Dr. R. S. Fox of Lander University. The staff of the Brevard College library, especially Brenda Spillman and Mike McCabe, assisted with locating reference materials. The staff at UNC Press, especially Mark Simpson-Vos and Mary Caviness, improved the overall quality of this book, and I have appreciated their effort and enjoyed working with them.

My husband, Ed Ruppert, has spent countless hours assisting me at every level: with scientific advice, editorial comments, suggestions for rewrites, photography, research, and species identification. Together we have seen, learned about, and discussed every organism described in this manuscript, and it is through his encouragement and support, both professional and personal, that I completed this project.

# Mountain Nature

# Introduction

## The Diverse Southern Appalachians

*Mountain Nature: A Seasonal Natural History of the Southern Appalachians* is a guide to the enjoyment and understanding of nature in the mountains of the Southern Appalachians. These gently rolling hills are carpeted with a rich and beautiful tapestry, intricately woven from diverse and colorful threads of biological diversity. Life flourishes in the temperate and moist conditions of the region, where ecology is elegantly complex, vistas are breathtaking, and waterfalls abound. This book takes root in these magnificent forests and will acquaint the reader with the seasonal appearance and activities of its natural residents. Residents and visitors to the Southern Appalachians, including its two national parks, eight national forests, and the popular Blue Ridge Parkway, will find this book to be a friendly and informative companion.

The Appalachian mountain range forms the backbone of the eastern United States. Imagine a giant sleeping on its side, facing west, with its head pillowed near the sea at the Gulf of St. Lawrence and its feet tucked into the earth in the foothills of Georgia and Alabama. Scientists usually divide the Northern from the Southern Appalachians at the limit of ice-age glaciation, which extended south to the northern border of Pennsylvania, but the plants and animals from central Pennsylvania and western Maryland are intermediate between northern and southern regions. For the purposes of this book, the northern border of the Southern Appalachian region is the Potomac River, which descends from West Virginia and flows along the border between Maryland and Virginia before reaching the sea. The eastern edge of the southernmost mountains is a steep drop-off, or escarpment, which rises dramatically from the Piedmont. The western edge plateaus and then descends gradually into river valleys that eventually join the mighty Mississippi River.

The Southern Appalachians, then, is the mountainous region extending from the Potomac to the northwestern corner of Alabama, and from the Blue Ridge Escarpment in the east to the Cumberland Plateau in the west. It is roughly six hundred miles long and encompasses about 35 million acres (see map). Of those 35 million acres, over 4 million (about 12 percent) are federal public land.

The outstanding features of the Southern Appalachians are their mountainous geography, the fact that they have been spared the historical effects of glacial scouring, their impressive age, the high rainfall they receive, and their high biodiversity. In addition, as noted above, a large percentage of the land they encompass is publicly owned. The highest mountains in the Southern Appalachians, the hips of the slumbering giant, are in the Great Smoky Mountains National Park and the surrounding counties of Tennessee and North Carolina. Mount Mitchell, at 6,684 feet, near Asheville, North Carolina, is the highest peak in the East. Farther north, in Shenandoah National Park, the mountains lose some of their rugged height, with the highest peaks at 3,000 to 4,049 feet.

Historically, the mountains were a barrier to human travel and colonization, resulting in the near isolation of their sparse human inhabitants as well as protection of the natural environment. While the fertile river valleys and the flatlands to the east and west of the mountain range were colonized first by native Americans and then by Europeans, only the most rugged individuals claimed the mountains proper. These mountaineers relied on the natural environment to sustain their lifestyle of subsistence farming and hunting, and their relative isolation contributed to a rich folklore and fierce independence. By the standards of the rest of the East and Midwest, towns were small, far apart, and focused on a few industries that were supported by natural resources, including mining and tanneries. During the age of the robber barons, from about 1870 to 1920, a few wealthy individuals purchased vast acreage and much of the timber was cut out of the region, except in the most remote and inaccessible areas. Even with these disturbances, the low population pressure and few industries allowed the natural environment to recover. In almost all cases, the current conditions of the region's forests and rivers are better than they were fifty to one hundred years ago.

Just as human cultural distinctiveness results from isolation, biological uniqueness arises when plants and animals are restricted over long periods of time to a specific local habitat. For example, an animal of limited mo-

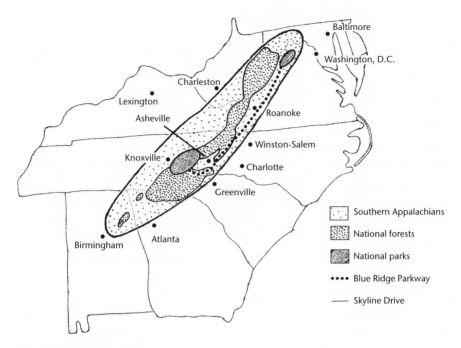

The Southern Appalachians

Key:
- ⣿ Southern Appalachians
- ⣿ National forests
- ⣿ National parks
- •••• Blue Ridge Parkway
- — Skyline Drive

bility, such as a small snail or salamander that is adapted to life in a wet, shady cove is unlikely to cross a high, dry ridge to reach the next cove. For the salamander and its offspring, the cove is an island surrounded by an impassable sea. Whatever unique biological traits appear in its family line will be inherited by future generations until all the descendants will be distinct from salamanders in other, equally isolated coves. Isolation and time, coupled with specialization to a particular habitat, lead to diversification.

A mosaic of habitats, each with its own unique natural community, resulted from the folding and upheaval of the earth's crust that produced the Appalachian Mountains. Some plants and animals thrive on the thin soils of windy ridge tops, while others flourish in the rich soils of protected coves. Plants that prosper on the sunny, dry, south-facing slopes of a mountainside are different from those on the shadier, moister, north-facing side of the very same mountain. Some organisms are specialized to live in the bold headwaters of a mountain stream that comes tumbling down a steep ridge while others prefer the slower-moving waters of large rivers. Uncommon habitats such as the highest peaks, high-elevation bogs, grassy balds, and rock outcrops each support a host of unique species. All these myriad habitats shelter an abundance of distinct species that are adapted to them.

But no one species is an island unto itself; within each habitat are many, interdependent organisms.

Life begets life, not only in terms of offspring, but also as each species, by its very presence, creates new conditions suitable for colonization by other species. A tree must be present before a resurrection fern can colonize its branches or before a northern parula warbler can forage for insects in its crown. Once the warbler is present, a black rat snake can eat its eggs and a mosquito or other parasite can drink its blood. As each new species arises in a natural community, it creates an opportunity, or niche, for other species to enter. Such diversification requires time: time for colonization, time for natural selection, and time undisturbed by major destructive events. When these conditions are satisfied, as in the Southern Appalachians, the result is a breathtaking richness and interdependence of living things.

For all practical purposes, the Appalachians are timeless. The Appalachian mountain range is the oldest on Earth. It has been heaved up and pleated by the colossal collision of continents, most recently 330 million years ago (see chapter 1). Since then, rain, ice, and wind have been eroding the uplifted rocks, creating rich and fertile soils in the valleys and wearing down the once high and jagged crests into the more rounded hills we see today. Although we still consider our 6,000-foot mountaintops to be high and rugged, they are a pale shadow of their original heights, estimated at 26,000 feet, or as high as the Himalayas today.

After eons of erosion and sediment accumulation in valleys and coves, cove soils are rich and deep. Plant life has flourished and diversified, for good soil encourages plant growth. Because soils vary in their mineral composition depending on the rocks from which they arose, different types of soils occur throughout the region and support a variety of plants. While some specialized plants grow in shale barrens, others grow in basic calcareous soils or in more acidic soils. Again, diversity of the underlying geology results in diversity of plants and the animals that depend on them.

Like some benevolent god, these immortal mountains have witnessed, supported, and nurtured a remarkable succession of life. When the Appalachians were young, some 300 million years ago, the dominant land animals were salamanders. There were no mammals, no birds, and no reptiles. There were also no flowering plants—no trilliums, no orchids, no oak trees, no maples. Forests were composed mainly of tree-sized ferns, horsetails, and clubmosses, whose diminutive descendants, relics of this past golden age, still grace the forest floor today. Not until around 150 million years ago did those prehistoric salamanders finally give way to the kings of the reptiles and dinosaurs roam the land. Herbivorous dinosaurs ate mostly conifers, tree ferns, and giant clubmosses, but early flowering plants, such as *Magnolia*, had arrived on the scene. About 65 million years ago, when the last dinosaurs went extinct, mammals and birds began to diversify in earnest, as did flowering plants and their insect pollinators. Throughout the entire history of mammals, birds, reptiles, and flowering plants, the Appalachians have been patiently protecting life. As long as there have been pink lady's slipper orchids, they have pursed their bright lips in Appalachian forests. When the very first wood thrush began to sing, that flute filled an Appalachian cove. Life's history runs deep in the fertile forests of the Southern Appalachians. No wonder we have so many different plants and animals living here! Salamanders, the most ancient group of land animals, are nowhere more diverse than right here in the Southern Appalachians.

During the vast expanse of time over which the Southern Appalachians have existed, they have avoided some of the major climate changes that affected the northern region, thereby preserving their biodiversity. For example, the southern mountains escaped the advancing glaciers of the ice age and the wholesale deforestation and valley scouring that accompanies glacial advance. The Pleistocene ice sheet extended southward as far as northern Pennsylvania in the Appalachians, and even farther south into the Ohio River valley. Over the last 2 million years, glaciers have expanded and contracted repeatedly, and the last cooling period ended only about 10,000 years ago. Each time they advanced, the glaciers scraped away vegetation in the northern region, as plants and animals moved southward. As climate warmed during periods between glaciations, plant seeds and mobile animals recolonized the northern and western regions from sources in the Southern Appalachian refuge. In fact, the Southern Appalachians owe some of their high diversity to the northern refugees who remained on the higher and cooler peaks of the southern mountains. Red spruce, balsam fir, the northern flying squirrel, and the saw-whet owl are all northern species that gained a foothold down south during cooler periods and then remained.

Because the Appalachian mountain range runs northeast to southwest, it not only serves as a highway for plant and animal migration during glacial advance and retreat, but it also tends to trap weather fronts moving from the southeast or from the northwest and cause them to drop their precipitation. The moisture-laden clouds that move up from the Gulf of Mexico or from the Atlantic Ocean (especially those associated with hurricanes) are intercepted by the highest mountains and trapped to the south of them. The clouds cannot rise and move over these mountains until they drop their ballast of heavy rain. The highest rainfall therefore occurs to the southeast of the highest elevations. The area with the highest recorded rainfall east of the soggy Pacific Northwest is Transylvania County, North Carolina, which *averages* an astonishing eighty-six inches of rain each year, an amount that nearly categorizes it as a rain forest.

Water is the substance of life. When water is abundant, life flourishes and diversity soars. Not only is water the chief component of living cells and bodies, but it is also the lifeblood of all living things. It circulates through plants and animals carrying oxygen, nutrients, and wastes, it supports body tissues, and it is a fundamental part of photosynthesis. Although significant rainfall occurs in all seasons of the year (see Table 1.1 in chapter 1), the rainiest season is summertime, when temperatures are highest, plant growth is lush, and the Southern Appalachians throb with life. Less rain falls in the autumn, when drier weather helps to produce the magnificent fall color that is legendary in the mountains. Even the pattern of rainfall, heaviest during the warm growing season, supports biodiversity.

While liquid water supports life, frozen water does not. The cold temperatures of winter, when water freezes into ice, limit the growth and distribution of many plants and animals. The tropics are so diverse because they never experience the majesty of snowfall and the beauty of ice palaces. We relinquish, as we must, a little diversity in exchange for the magnificence of winter. Yet these mountains are southern after all and are not encumbered for long by the burden of ice.

All that rain must go somewhere, and what is not absorbed by plants goes rapidly downhill. The combination of high rainfall and mountainous topography creates the "Land of Waterfalls" in Transylvania County, where tiny mountain streams dance and slide over faces of stone and larger rivers roar and tumble over rocky ledges in dramatic veils of water. The complex forms of these mountain streams and rivers, their saturation with oxygen from the tumbling action of the waterfalls, and the rocks, pools, and sandy stretches they contain create manifold habitats for myriad aquatic animals. For example, the cool, clear, oxygenated waters of the Southern Appalachians contain more kinds of freshwater fish than in any other temperate area of the world. The greatest freshwater mussel diversity also occurs right here. Unfortunately, freshwater mussels, which live in rivers and filter the water for their food, are among the most imperiled animals on Earth. Of the eighty-five local species, fully half have either already gone extinct or are listed as endangered species. Pollution of the rivers by toxins and silt, coupled with the historical exploitation of mother-of-pearl for buttons and later as seeding chips for cultured pearls, caused their decline.

Timeless mountains spawn ancient watercourses, the arteries and capillaries that collect and distribute water to form a wellspring of life. Two of the three oldest rivers in the world are in the Southern Appalachians. Both the French Broad River and the ironically named New River are archaic rivers, contemporary with the equally ancient Nile, and they harbor relic species. Among the most primitive of freshwater fishes is the lake sturgeon, which is the current subject of a restocking effort in the French Broad River.

The Southern Appalachians, encompassing the highest region of the Appalachian mountain range, support the highest biodiversity, which reaches its peak in the Great Smoky Mountains National Park (GSMNP). The GSMNP, like Australia's Great Barrier Reef and other well-known biological wonders, has been recognized as an International Biosphere Preserve by the United Nations. Currently, the GSMNP is hosting a research program called the All Taxa Biodiversity Inventory, whose purpose is to document the park's biodiversity. Over 12,000 species have already been identified within the GSMNP boundaries, but there are estimates of another 90,000 or so that remain to be found and described.

The Southern Appalachians offer boundless sensual pleasures of pattern, color, sound, and motion that change with the changing of the seasons. Autumn's hillsides of patchwork plaid are replaced by the silence and majesty of winter's snow-cushioned forests and ice-stilled streams. In spring, the white blossoms of the silverbell tree fall like springtime snow as they give way to the profoundly deep scarlet of summer's cardinal flower. The basso profondo of a great horned owl is exchanged for the coloratura soprano of a winter wren, which is itself replaced by countertenor voices of tree frogs as each season cycles to the next. Whether it is the fluid grace of a river otter or the slick water that falls as foam and mist over primordial rocks, these and countless other pleasures are the inspiration for this book. But so too is the knowledge of nature's interconnections and cycles, which elicit a deeper appreciation of its beauty. Everywhere there is relationship and rhythm, waiting to be seen and heard, not only by scientists or philosophers, but by all nature enthusiasts through observation and discovery. As the poet A. R. Ammons writes in his poem "Sphere," "Science . . . makes mysticism discussable." *Mountain Nature* is an entrée to this conceptual level of understanding. For me, the Southern Appalachians are beautiful as both art and science, and in this mountain region, the two realms are deeply and wonderfully connected. With *Mountain Nature*, I hope to encourage you to share in this sense of discovery and wonder.

# The Nature of Cycles

1

Nature animates itself in cycles. Some cycles, such as sleep and wakefulness or hunger and satiation, are of immediate concern. Others, such as the transition from day to night, or from spring to summer, to fall, and then to winter, are broader, encompassing other cycles. Because natural cycles occur at all scales of observation, from the motions of celestial objects to those of infinitesimal particles, they may seem too vast or too insignificant for us to comprehend, yet all these different natural rhythms combine harmoniously, like some vast musical score, to provide order, vitality, and charm to our world.

In the Southern Appalachians, the seasons constitute the most conspicuous and important natural cycle. The muted tones of winter swell into the melodious spring, settle into the deep harmonies of summer, and then, after the colorful fall cadenza, diminish into the pianissimo of winter. We eagerly anticipate the changes from winter snows to spring flowers, then to summer greenery and autumn leaves. Like us, Appalachian animals and plants also sense seasonal progression and prepare for the changes in form and tempo. Unlike most of us, however, they are fully engaged in each season's performance, resonating with each note and responding as gently plucked strings.

Within that yearly cycle of the seasons, from spring to winter and back again, is an underlying daily rhythm, pulsing like a throbbing heart. The daily cycle, from dawn to dusk and then back to dawn again, is twenty-four hours long. After 365 daily beats, the yearly cycle has completed an entire revolution. Our bodies, as well as those of other living organisms, track and respond to these cycles. We have a biological clock as well as a biological calendar deep in our core.

Sleeping and waking, growing and shedding underfur or leaves, and migrating across continents are all cyclic patterns in nature. We know to expect an increase in bat and moth activity at night and to watch for waves of monarch butterflies in the first cool days of fall, just as the leaves of deciduous trees begin to change color, because the shift from day to night or summer to fall causes animals and plants to respond predictably. Long-term events, such as periods of mountain building or the advance and retreat of ice sheets, are also cyclic and encompass seasonal and daily rhythms. Such geological cycles have a profound influence on the composition of natural communities of living things. Daily, seasonal, and long-term cycles of change provide the organizational framework of this book, but the range of nested rhythms, like Emersonian circles, extends from the atomic to the astronomic and provides an assuring regularity to the music of the spheres.

## Daily Cycles and Biological Clocks

Our moods, appetites, and energy levels change over the course of a day as our biological clock ticks. Most of us are asleep at 3 A.M. but awake at 9 A.M. We become hungry at about the same time each day. Our body temperature and blood pressure are lowest at night and highest during the day. These daily cycles (also called circadian rhythms) are generated by an innate biological clock. Even in the absence of environmental cues as when, for example, bean plants are kept in constant light or rats are kept in constant darkness, we still respond as if the sun is rising and setting. All organisms that have been studied, from bacteria to humans, have biological clocks that control recurring cycles of processes such as cell division, hunger, wakefulness, blood pressure, and body temperature. From the outset, we've all got rhythm!

Circadian rhythms, however, can be adjusted by cues from the environment. This flexibility allows environmental information, such as the shift in day length as winter becomes spring, to reset the clock. For example, a circadian rhythm is the cause of jet lag—when you suddenly arrive in a different time zone, your clock remains set to the old time and causes you to become sleepy at inappropriate times, but over the course of a few days, your biological clock is reset to the new time and you feel better, usually just in time for your return trip back to another time zone! In mammals such as ourselves, the clock is composed of a group of cells lodged in the hypothalamus of our brain. The hypothalamus, part of the forebrain, controls the secretion of many hormones that regulate behavior and biological processes.

## The Daily Cycle of Sleep in Animals

The most obvious daily cycle in animals is the cycle of sleep and wakefulness. Diurnal animals are active during daytime and sleep during night, whereas nocturnal animals are active at night and sleep during daytime. But why do animals sleep? Wouldn't it be advantageous to be awake all the time, constantly on watch for predators, food, or mates and able to respond immediately should one appear? Sleeping would seem to put an organism at a disadvantage, yet animals still sleep. There must be something to gain!

Perhaps the most general function of sleep is to conserve energy by reducing metabolism. In particular, small animals such as hummingbirds, shrews, and bats have very high metabolic rates that require huge amounts of food to sustain. While they are awake, they eat more or less constantly. Starvation is a very real threat to these small furnaces. Dead but uninjured shrews (see chapter 5) most likely died of starvation, especially in early spring, when nights are cold and food is scarce.

Hummingbirds (see chapter 2) feed during daylight only, when they can see the flowers and insects that provide them with high-energy food. And feed they must! The adult animal with the highest recorded metabolic rate ($85$ ml $O_2$/g hr) is a hovering Allen's hummingbird. Our Appalachian species, the ruby-throated hummingbird (*Archilochus colubris*), is no less impressive. When at rest, its heart beats 615 times and its lungs draw 250 breaths each minute. When flying, its wings beat seventy times each *second* and its heart rate doubles to around 1,200 beats per minute. Human metabolism is about twenty times lower.

At night, hummingbirds cannot feed and therefore cannot fuel that incredible metabolism, especially during periods in early spring when cold weather is coupled with few nectar-providing flowers. Instead, they have a remarkable adaptation called daily torpor, which is the equivalent of seasonal hibernation, a good example of the similarities that exist between daily cycles and seasonal ones. Both daily torpor and seasonal hibernation are controlled by biological clocks. During torpor, a bird's fuel consumption drops, as measured by the amount of oxygen consumed for combustion, by about a hundred times. Then, just before dawn, its metabolism increases and its body temperature returns to normal as it arises, like Lazarus, from what seems to be a near-death state.

## Daily Cycle of Sleep in Plants?

Plants don't sleep — or do they? Most plants, just like animals, have circadian rhythms and make cyclic changes over the twenty-four-hour day. For example, most plants shut down their photosynthetic machinery or close their stomata, which are openings for gas exchange on their leaves, on a daily cycle. Because photosynthesis can occur only during daylight, the plant conserves energy by reducing its metabolism at night. Since both plants and animals conserve energy in this same way, most scientists concur that plants do, indeed, sleep, even though we more often associate sleeping with animals.

Some plants even undergo so-called sleep movements, a topic that so intrigued Charles Darwin that he devoted an entire chapter of his book *The Power of Movement in Plants* to the subject. Many species, especially those in the pea family, have circadian leaf movements, in which the normally horizontal leaves are folded vertically at night. If the plants are constrained so that they are unable to fold their leaves at night, then their growth rate declines. Mimosa (Figure 1-1), wood sorrels, and bean plants all exhibit sleep movements. Incidentally, the eastern sensitive-brier and other plants that fold their leaves in response to a physical disturbance, such as an animal brushing against them, use a similar mechanism to fold leaves, but the stimulus is physical, not circadian.

## Communication at Night

Since we are primarily visual animals adapted for daylight, nighttime confronts us with challenges that can be downright scary. Vision, our major sense, is compromised by low levels of light. For nocturnal animals, nonvisual senses such as touch, hearing, and smell dominate sight in gathering information about the environment. You may have noticed in yourself a heightened sensitivity to sounds, odor, and touch during nighttime walks in field or forest. A gentle rustle of wind through leaves can sound like an elephant charging through underbrush; the delicate scent of flowers can be stronger than perfume; and the slightest breath of wind may stir the hairs on the back of your neck.

Most nocturnal organisms have well-developed senses of hearing and touch. Nighttime is full of booming frog voices, chirping crickets, buzzing katydids, whining mosquitoes, and the screams of bats (though unheard to us). These sounds, specific to each species, not only allow members to locate each other but also assist in selection of a mate. A female cricket is more impressed by a male that sounds good than one who looks good!

FIGURE 1-1

Mimosa (*Albizia julibrissin*) leaflets are open during the day (left) and closed at night.

At its most simple, a hearing organ consists of a taut membrane stretched over a rim, like a drumhead. Physical disturbance of the air causes the drumhead (our eardrum) to vibrate, and those vibrations trigger nervous impulses that the brain interprets as sound. Each drumhead is called a tympanum, from the same word root as tympani, the drums of an orchestra. In frogs, which communicate with sound, the tympanums are large round disks, one located on each side of the head (Figure 1-2). Insects that communicate using sound may also have a pair of tympanums, but they are usually located on the legs (Figure 1-3). Other insects lack a tympanum but have specialized hairs that are tuned to vibrate at a particular pitch, like tuning forks. Male mosquitoes can recognize females of their species and some caterpillars can detect wasp predators because of sensory hairs that are tuned to the wing beat frequency of the female or the wasp, respectively. It must be an electrifying sensation as these surface hairs all buzz in response to an opportunity for either sex or predation! Echolocation is a specialized sense of hearing that has been well developed by bats (see chapter 4), and many of bats' oddities are really just adaptations to this sense.

The senses of smell and taste are also well developed in nocturnal organ-

FIGURE 1-2

The wood frog (*Rana sylvatica*), like other frogs, possesses a tympanum or external eardrum, which is a circular disk just below and behind the eye. Unlike most other frogs, it lays its eggs in winter months.

isms. For instance, dogs have a much better sense of smell than humans and are a thousand times more sensitive to odors. While domestic dogs may not be primarily nocturnal, they are often active at night, as are wolves, coyotes, and foxes, their close relatives. Mammalian noses are particularly receptive to the molecule butyl mercaptan, which is found in mammalian sweat glands. A skunk's attention-grabbing and enduring odor arises from a concentrated broth of butyl mercaptan that is produced from modified sweat glands (see chapter 5).

The humid air of warm summer nights carries and disperses odor molecules more widely than the dry air of daytime, resulting in the nocturnal fragrances familiar to poets and lovers. Mammalian noses, however, are not the only organs that detect airborne chemicals. Snakes (see chapter 3) use their tongues to detect the chemical signals emanating from their prey. Each flicker of the tongue gathers odor molecules from the air and deposits them on a sensory organ in the roof of the mouth. Some insects, such as moths, use their antennae to detect odors, and males typically have larger antennae than females because males locate females by their subtle but alluring scent

FIGURE 1-3

This southeastern field cricket (*Gryllus rubens*) and other insects use a tympanum that is located on the tibia of each foreleg to transduce sound. The left front leg (inset) is magnified to show the tympanum, which appears as a light oval. This female cricket uses her long, thin ovipositor to deposit eggs deep into soil.

(see chapter 3). Insects also have chemical detectors on their legs. Houseflies, for example, taste food by walking on it.

Some nocturnal organisms rely on sight as their dominant sense and have adapted to the low light levels of night by improving their vision. As a general rule, these animals have large eyes that are extremely sensitive to light and often have a reflective layer. Many mammals that are active at night, such as opossums, have such sensitive eyes that the bright light of day disturbs them (see chapter 5). Some nocturnal animals, such as fireflies (see chapter 3), produce their own light and respond to particular patterns, colors, and frequencies of light emanating from other individuals.

## Seasonal Cycles and Biological Calendars

The Southern Appalachian year is divided neatly into three months each of winter, spring, summer, and fall. December, January, and February make up winter; spring is composed of March, April, and May; summer has June, July, and August; and September, October, and November are fall months. Each season is long enough for us to experience its characteristic features of temperature, precipitation, day length, and the growth of plants and animals, but short enough for us to sense and appreciate the gradual change to the next season. Just as we begin to tire of winter's cold, a gentle warm rain thaws the ground and releases wood frogs, a sure sign of spring's approach (see chapter 5). Similarly, the fatiguing heat of late summer is soon relieved by the first returning cold front, as monarch butterflies follow the weather fronts south on their fall migration (see chapter 4).

Each season is characterized by a unique combination of day length and temperature (Table 1-1). Since the shortest day and longest night of the year fall on the first day of winter (usually December 21), the winter season is distinguished by short (but increasing) day lengths as well as by cold temperatures. The length of the night is longer than that of the day during winter, and it is these long, cold nights that help keep environmental temperatures low. Spring arrives on March 20 or 21, when the length of the day finally equals that of the night. Spring is the season of increasing day lengths and temperatures. Summer begins on the longest day of the year (usually June 21), which makes it the season of long (but decreasing) day lengths as well as warm temperatures. Those long, sunny days allow the temperature to reach its peak. Fall usually arrives on September 22 or 23, when night and day lengths are once again equal. As fall progresses, day lengths decrease, nights grow longer, and temperatures fall. During the fall season, long nights allow the environment to cool more than it heats during the shorter day, as the season cycles back toward winter.

One of the functions of our calendar is to keep track of the seasons, so we know when to plant and when to expect harvest or the migration of animals. Agricultural societies commonly used stone and wooden structures as calendars to track the position of the sun in the sky over the course of a year. Like us, other organisms track the yearly calendar to predict favorable times for food availability and reproduction. For most animals, offspring are born during the spring season to coincide with abundant food and warm temperatures, with enough time to mature before winter arrives. Similarly, plants must bloom, ripen fruit, and set seed (reproduce) while they have enough light (the energy used to produce food) to accomplish these tasks.

While temperature affects the rate of growth and development of an organism, it is not the only factor that determines when an organism will grow or reproduce. Temperature, after all, fluctuates from day to day during any season. In the Southern Appalachians a 70° day can occur in winter, spring, summer, or fall, and that day can then be followed by another that is significantly warmer or cooler.

Temperate plants and animals have adapted to avoid being misled by

TABLE 1-1

Climatic conditions (averages) in Great Smoky Mountains National Park
at Gatlinburg, Tennessee (1,454 feet elevation), 1971–2000

| SEASON | MONTH | HIGH (°F) | LOW (°F) | PRECIPITATION (INCHES) [DAYS >0.1 IN.] | SNOW (INCHES) | DAY LENGTH (HOURS) |
|--------|-------|-----------|----------|----------------------------------------|---------------|--------------------|
| Winter | December | 51 | 28 | 4.5 [9] | 0.7 | 10 |
| | January | 48 | 25 | 4.9 [9] | 3.6 | 10 |
| | February | 52 | 26 | 4.3 [8] | 1.4 | 11 |
| Spring | March | 61 | 33 | 5.6 [10] | 1.2 | 12 |
| | April | 69 | 39 | 4.4 [8] | 0.9 | 13 |
| | May | 76 | 49 | 5.6 [10] | 0 | 14 |
| Summer | June | 82 | 57 | 5.8 [10] | 0 | 14.5 |
| | July | 85 | 60 | 6.0 [10] | 0 | 14 |
| | August | 84 | 60 | 4.6 [8] | 0 | 13.5 |
| Fall | September | 79 | 54 | 4.6 [7] | 0 | 12.5 |
| | October | 70 | 42 | 3.0 [5] | 0 | 11.5 |
| | November | 60 | 33 | 4.0 [8] | 0.2 | 10.5 |

*Source*: National Climatic Data Center, Asheville, N.C.

occasional warm temperatures during the cold months of the year. Many
plants, such as most temperate deciduous trees, require a minimum number
of days above a certain temperature before they will leaf out. Still, they can
be fooled by temperature. Several days of unseasonably warm weather in
early spring, for instance, can accelerate the emergence of flowers or leaves.
When cold temperatures return, the young leaves may be killed, resulting in
wasted energy for the plant. Most animals base their biological calendars on
day length rather than temperature because the length of day changes in a
predictable manner: if day length is increasing each day, and if day length
exceeds night length, then it must be spring.

## Day Length as the Cue for Seasonal Calendars

How do animals determine day length? Mammals and birds use their eyes
and the pineal gland of the brain. The pineal secretes the hormone melatonin,
but only at night. During periods of longer nights (and shorter day lengths),
therefore, birds and mammals produce more melatonin. High levels of melatonin block the release of hormones that control growth and reproduction

FIGURE 1-4

In the Carolina anole (*Anolis carolinensis*) and other lizards, the "third eye" or pineal gland is protected by a large, modified scale centrally located between the eyes on the top of the head.

from the hypothalamus—the region associated with the biological clock. Day length, therefore, affects the pineal gland, which in turn affects an organism's biological clock and calendar. The pineal gland of mammals and birds is descended from the pineal eye of fishes, amphibians, and reptiles. In most of these animals, this single eye, complete with retina, sits on the surface of the brain and is permanently directed skyward. (Is it any surprise that Descartes believed the pineal gland to be the seat of the soul?) In anoles, such as the Carolina anole, the skull and scales overlying this third eye are modified and translucent (Figure 1-4). In other animals, such as iguanas and fence lizards (see chapter 3), the pineal eye is located beneath a hole on top of the skull, like an extra eye socket. Animals with a functional third eye use it to determine day length. Because the pineal *gland* of mammals and birds (and a few reptiles) is no longer sensitive to light like the pineal *eye* of more primitive animals, special nerves connect it to the paired eyes, which provide the day length information.

Plants also use light as a cue to produce hormones that regulate their growth. Lacking brains and nervous systems, however, plants produce their hormones in various tissues, such as shoot tips, root tips, or developing fruits, and transport the hormones through the rest of the plant body in their tubular vessels, the xylem and phloem. Charles Darwin, with the help of his son Francis, performed some of the first experiments that demonstrated how a hormone produced in one region of the plant caused a growth response in another part, and that light was the cue used. They found that seedlings normally bend toward light as they grow. When the tip of the seedling is covered with an opaque tube that blocks light, however, the seedling does not bend; when it's covered with a transparent tube, it does. Scientists have since discovered that an auxin hormone is produced by the illuminated leaf tip and transported throughout the plant. The hormone inhibits cell growth on the illuminated side of the shoot, causing the growing seedling to bend.

The seasonal blooming period of many plants is determined by day length. Chrysanthemums, poinsettias, and other short-day plants bloom in the fall, when the days are shorter than the night. Iris, lettuce, and other long-day plants bloom in the spring, when the days are longer than the night. For these plants, a flash of artificial light during the night is enough to turn one long night into two short ones, thus changing the blooming period, a trick that horticulturalists can use to produce flowers at any season. Some plants, however, do not rely on photoperiod. They bloom as soon as they mature, regardless of day length. Examples of these day-neutral plants include tomatoes and dandelions.

## Tree Rings: Evidence of Seasonal Cycles in Plants

Tree rings, the alternating bands of dark and light tissue visible in the cut stump of a tree, record the change of seasons. Each ring is composed of both a dark and a light band, which together constitute a year's growth. The lighter band is made of large, water-conducting, xylem cells formed during spring and summer, the period of most active growth. The darker ring consists of small xylem cells that form during early fall, as growth slows. These smaller diameter cells conduct less water but add more support to the tree. Because the cells are smaller and closer together, the band appears darker than that formed during the spring and summer.

Xylem, which forms the wood of the tree, is the major tissue of water conduction and physical support in all plants. The reason trees can survive even when their center, called the heartwood, has rotted is because the center is the oldest and least functional wood; the xylem vessels are mostly plugged up with resin and no longer conduct water. By contrast, the newest and most active xylem is outermost. This active xylem is called sapwood because the flow of sap is greatest there.

The other major conducting tissue, the phloem, is a thin layer just underneath the bark, which transports food from the leaves to other organs such as roots and developing fruits and, in spring, moves sugars up from the roots to the branches and developing leaves. Trees can be killed by shallowly girdling the tree around its circumference, which destroys the ability of the phloem tubes to transport nutrients. Phloem is also produced every year, but the older layers slough off as bark.

## Migration and Hibernation: Seasonal Cycles in Animals

Because the Southern Appalachians experience all four seasons in equal measure, residents must be responsive to the melody of each season. For animals, as for plants, the cold temperatures and low food supply of winter typically test their endurance and limit their success. Some animals avoid the difficulty of winter by migrating to warmer climates, and others "sleep it off" by hibernating. Changes in day length trigger these cyclical responses.

Migration, the seasonal movement of a population over long distances and to different habitats, is a behavioral adaptation to predictably changing conditions. For example, warblers migrate northward in spring to exploit the spring explosion of insects (see chapter 2), but they move south in winter as their northern food supplies dwindle. Migration is possible only for animals that are equipped to travel long distances. Many species of birds migrate, as do a few species of bats, since flight permits long-range travel. Only a few winged insects migrate; most insects have other ways to adapt to changing and adverse seasonal conditions. Some large mammals also migrate, but none of these live in the Southern Appalachians.

Hibernation is another adaptation to winter and most commonly occurs in medium-sized mammals. Hibernation is a physiological adaptation in which the animal is able to reduce its body temperature, slow its metabolism, and survive off its fat reserve during periods of low temperatures and low food availability. Hibernation is a response to a seasonal shift, whereas torpor and sleep are responses to daily cycles.

By slowing its metabolism, a groundhog, or woodchuck (*Marmota monax*), can spend winter inside its burrow without eating, drinking, or awakening. Its heart and breathing rates slow down, the amount of oxygen used is reduced (which is the most common measure of metabolism), and body temperature drops to nearly that of the air temperature of its burrow, which is about 50°F. Similarly, most Appalachian bats also hibernate, with body temperature and metabolism dramatically reduced. By contrast, black bears

Both temperature and day length influence the life cycle of animals and plants. Most often, day length is the cue used to prepare for advance of the seasons. Temperature affects the rate of chemical reactions in the body and the speed of development, growth, and metabolism. Global warming raises the average yearly temperature of the Earth but does not change day length. Some animals may be more affected by temperature change than others. For example, warmer temperature results in earlier hatching dates for caterpillars and accelerates their development. These caterpillars are the main food source for migratory songbirds and their chicks. Most migrating birds, however, time their spring migration by day length. Thus, the birds are arriving north to find their food supply of caterpillars diminished because they have already transformed into adult moths. In turn, the number of chicks the birds can successfully rear declines. This trend has been best documented in European migratory songbirds but is expected to affect any migratory bird that feeds primarily on insects.

(*Ursus americanus*) experience a much less intense type of hibernation, sometimes referred to as seasonal torpor. Bears do not eat, drink, urinate, or defecate while in their winter den, but their metabolism slows only slightly and they can awaken easily. Bear cubs are born while the female is denning for winter, and she nurses them for several weeks before leaving the den in the spring. Eastern chipmunks (*Tamias striatus*) spend short periods in a deep sleep but then awaken and feed on their stored cache of nuts, occasionally even leaving their winter den.

Reptiles and amphibians of temperate areas are sometimes described as hibernating because they sleep away the winter months in protected dens. Since their body temperatures passively track those of their environment, little metabolic change is required. Most snakes enter winter dens, most turtles bury themselves in the soft muck of pond bottoms, and most frogs and salamanders bury themselves underground.

Of course, many animals neither migrate nor hibernate but remain active during the winter. These include species that are too small to store up enough fat to survive hibernation or to travel long distances and large animals that are protected from cold by virtue of their body size. Many winter residents, such as mice (see chapter 5), are generalist omnivores, able to consume many different types of food, ranging from seeds to insect pupae. Strict herbivores, such as white-tailed deer, may switch from nutritious herbs and grasses to twigs and bark. Prey is available year-round to carnivores and scavengers, but that prey is probably less abundant in winter because of hibernation and migration and because fewer naïve young animals are about.

## Seasonal Cycles of Aquatic Animals

Even aquatic animals, which are buffered from extremes of temperature by water, exhibit seasonal cycles. In the lakes and ponds of the temperate zones, especially in North America and Europe, biologists have tracked the cyclic change of plankton. Plankton is composed of single-celled protozoans and algae (together called protists) as well as tiny multicellular animals. Plankters float and swim in the water but are too small to swim against a current, which distinguishes them from larger aquatic animals such as fish. Scoop up a glass full of lake or pond water, hold it to the light, and tiny moving specks will be visible. Put these plankters under a microscope, and their swimming legs, protective shells, and pigmented eyes stand out. Over the course of the seasons, different species come to dominate the plankton in a predictable sequence.

Early in the spring, when water temperatures are uniformly cold from top to bottom, strong winds can mix the water, stirring up debris and nutrients from the bottom and adding oxygen from the surface. The resulting rich soup is perfect for the spring plankton bloom that gets under way with warming temperatures. Spring ponds quickly change from glass-clear to greenish as the water thickens with the living bodies of myriad plankters. In the spring, small, fast-moving species are common in plankton. Tiny photosynthetic algae thrive because the water is relatively clear, allowing sunlight to penetrate into the depths, and few of their predators are present.

During the summer months, the upper surface waters warm but the lower waters stay cold. The warm upper waters float on the denser, cold water, essentially dividing the lake into two layers. (When I swim in local summer ponds, my feet are cold while my shoulders are warm.) Nutrients, in the form of dead organisms, fall to the bottom, but oxygen enters from the top. As the lake becomes depleted of oxygen at depth, animals are forced up into the thin upper layer, and predators join their prey. As predation pressure increases, larger plankters that have numerous defensive spines come to dominate the plankton (Figure 1-5). Many of the photosynthetic species disappear, eaten by the numerous predatory species and stressed by the low level of nutrients. Summer ponds are typically thick with plankton, but only in the upper, warmer layer.

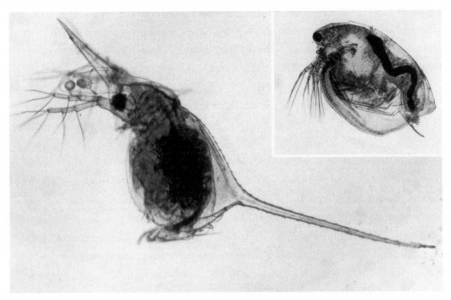

FIGURE 1-5

When predation pressure is highest, planktonic animals that are spiked or otherwise protected, such as the introduced non-native *Daphnia lumholtzi* (left), are more common than their less protected relatives, such as *Ceriodaphnia lacustris* (right).

Fall arrives with cooler temperatures and winds. Once the surface waters have cooled to a temperature close to that at the bottom, the lake is no longer stratified and is easily mixed by wind. The mixing creates a second plankton bloom in the fall season as bottom nutrients are again made available and as predators disperse throughout the lake. For a few weeks, before winter sets in, the small, photosynthetic plankters again dominate the plankton. Local ponds continue to mix throughout the fall and winter as long as they remain ice-free.

Winter ponds are clear as glass because few plankters are present to cloud the water. During winter months, most of the adult plankters die, but their eggs or resting stages overwinter on the bottom. They hatch in the warming days of spring to begin the cycle anew.

## Appalachian Trout and Other Fish

Like the plankton, fish also undergo seasonal cycles of reproduction and growth. Because most Appalachian streams are small and shaded, their temperatures and oxygen content vary less over time than that of ponds. The physical action of water tumbling over rocks introduces air into the water, and most of our streams are nearly saturated with oxygen. In addition, cold water is able to hold more dissolved gas. Not surprisingly, animals that live in these streams are very sensitive to decreases in oxygen because they rarely experience that problem. Trout, darters, and shiners (stream minnows) die quickly when the oxygen level drops.

By contrast, the surface of ponds and lakes are usually exposed to sunlight and absorb heat quickly during the summer. Surface waters become warm and float on the denser, cooler deep waters, effectively stratifying the lake and preventing oxygen from entering the lower layer. In the winter, because the water in ponds is still, it freezes and forms a barrier between the unfrozen water and the air. If the barrier persists long enough, oxygen can become depleted. Fish that live in ponds must adapt to low oxygen conditions, and few native Appalachian fish can. Most ponds and reservoirs in the Appalachians were built during the twentieth century and are stocked with fish, such as largemouth bass (*Micropterus salmoides*) and bluegill (*Lepomis macrochirus*), that normally occur in warm-water ponds of the coastal plain.

Because water temperatures during late summer are the highest of the year, and warm water holds less oxygen than cold, late summer is typically the most stressful season for aquatic animals. Under crowded conditions, such as those in trout farms, the water is usually supplemented with pure oxygen, which is pumped in from a tank of the liquefied gas, or the fish will not survive. Fall is a more favorable season and is the season that native brook trout (*Salvelinus fontinalis*) as well as introduced brown trout (*Salmo trutta*) reproduce. During the fall and winter, stream water is highly oxygenated. Trout eggs and fry, which are even more sensitive to oxygen levels than adults, thus have adequate oxygen for their development. In the spring, as more food in the form of insect larvae becomes available, the small fry hatch and are waiting for them. Thus the reproduction of trout is related to environmental temperature and to the availability of food, both of which cycle seasonally.

The Appalachians host three species of trout, but only one is native. Brook trout, also known as speckled trout, are mostly restricted to the upper reaches of undisturbed streams, where they have been pushed by competition from the two introduced species. Rainbow trout (*Oncorhynchus mykiss*) are native to the western states but were introduced in the East around 1880. They are aggressive, fierce competitors, and fishermen enjoy catching them for that reason. Brown trout were introduced from Europe around 1905. Both species were introduced to "improve" the fishing, for the native brookies are easy to catch and therefore quickly depleted. Unfortunately, the non-native,

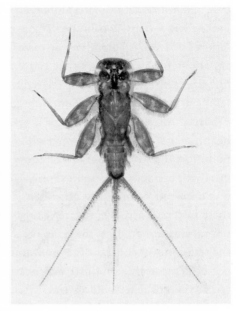

FIGURE 1-6

When mayfly adults (left) emerge, trout and other predators feast on them, but the adult form lives only a day or two. The wingless larva (right) is aquatic and long-lived.

invasive species displace the native trout and reduce their numbers, as well as those of many other animals and plants.

## Trout Food: Aquatic Insects

All three species of trout feed on aquatic insects, primarily the larvae or nymphs of mayflies, stoneflies, caddisflies, dragonflies, damselflies, craneflies, and midges. As adults, these insects all fly, as is suggested by their common names, but as eggs and larvae, they live under water (Figure 1-6). In fact, most of their lives are spent as larvae. The adult form may not even feed, relying on food acquired as a larva to sustain it as an adult while it finds a mate and lays eggs for the next generation. As larvae, most of these aquatic insects feed on the leaves that fall into streams. Some grind up the leaves themselves, and others scrape off the film of bacteria and fungi that grows on the leaf surface. It may take more than one year for the larvae to grow large enough to metamorphose into adults.

Most aquatic insect emergences take place in the spring or summer and involve huge numbers of individuals. Mayflies, for instance, are legendary for their massive emergences. Hundreds to millions, depending on the species, of adults emerge, mate, and lay eggs all in one or two days, rarely living more than seventy-two hours. Their group name, Ephemeroptera, means "short-lived wings," for the vast majority of their lives is lived as wingless aquatic nymphs. The cues that cause emergence are not well understood, but emergences are definitely related to the seasons. Some combination of temperature and day length undoubtedly play a role.

Surveys of aquatic insects are a standard method for determining the health of a stream. Stoneflies, mayflies, and caddisflies are very sensitive to pollution. If a stream contains all these insects, we know it is in good health, but if, for instance, it contained only caddisflies, we would assume its health was impaired because the other sensitive species were absent. Monitoring these insects is actually a better gauge than just testing the water itself, for several reasons. For one, these insects live in the stream year-round, whereas the water moves constantly. A pollutant might be washed downstream and not be detectable in an upstream location, but dead or absent indicator insects would be detectable anytime after the event. For another, they are easy to collect and identify, whereas testing water for possible pollutants can be complex and expensive. Finally, research has shown that some species are more sensitive than others, so a graded index of water quality can be constructed for the stream based on its insect community.

## Longer-term, Multi-year Cycles

Although daily and yearly cycles are the easiest cycles to visualize, longer-term cycles are equally important to the success of a population. Some of these long-term cycles revolve around changes in the reproduction of populations of animals and plants. Some years are better than others for reproduction and growth in a variety of organisms. When there is a good crop of youthful recruits, populations increase and may even expand beyond typical boundaries, thus entering new territories. During years when environmental conditions are unfavorable, populations may retract into smaller areas and become isolated from each other. These shifts affect the long-term viability of the species, its range, and its propensity for forming new species, for isolation encourages speciation (see Introduction).

A second important long-term cycle involves climatic change caused by geological shifts. These much longer cycles are related to the uplift and erosion of the Appalachian Mountain Range and the movement of continents over the surface of the Earth. Because the Appalachian Mountains have been present on Earth for such an immense span of time, the organisms that call

Most rivers in the Appalachians are stocked with trout to keep up with demand. During the trout-fishing season in North Carolina, for example, which begins on the first Saturday in April and ends in the fall, the Wildlife Resources Commission stocks the state's rivers with over 600,000 trout. Forty percent of them are brook, 40 percent are rainbow, and 20 percent are brown trout. While brook trout are native, they are not nearly as aggressive as rainbows, and it is the rainbows that most people favor. Brown trout are wary and difficult to catch, and some anglers prefer that challenge.

them home have faced the effects of ice-age advances and retreat. They have seen periods of time when conditions were nearly tropical and other times, millions of years later, when conditions were colder than they are now. These long-term cycles thus change the biodiversity of the region.

## Tree Reproduction and Long-term Masting Cycles

Although the cycles of each day and season are predictable, each one is not exactly like the next. Fluctuations in temperature and precipitation within a cycle can affect organisms that rely on day length as a cue to the season. Day length may indicate that March 20th has arrived, and organisms respond to spring by gearing up for reproduction. However, conditions can vary from one spring to the next. A particular spring may be warm or cool, wet or dry, with a late frost or no frost at all, and these environmental variations necessarily influence an organism's growth and reproduction.

In some years, environmental conditions are so favorable that trees are able to reproduce heavily. The trees have enough water to convert every leaf and flower bud into viable leaves and flowers; warmth encourages a booming population of insect pollinators; gentle winds and dry conditions help distribute pollen far and wide; warm and sunny weather supports production of food through photosynthesis. When these optimal conditions combine, huge crops of seeds are produced, a phenomenon known as masting. Trees that produce heavily in one year, however, may require several years to recover from such a heavy reproductive effort, because so much of their energy went into seeds rather than their body tissues. Heavy mast years are often followed by years of low (or no) seed production.

In mast years, there are so many seeds produced that not all of them can be eaten by animals, and by "flooding the market," more seedlings result. In a mast year, whole populations mast, not just individual trees. For example, every eastern white pine in the forest may produce a heavy crop of seeds during a mast year, but the next year, the same population will produce very few. Many forests that are naturally dominated by (not planted with) white pine have trees of the same age because the seeds all appeared in a mast year.

Masting has been best documented in forest trees. In addition to white pine, American beech, firs, spruces, hickories, and oaks are masting species of the Appalachians. Other trees, such as most maples, produce about the same number of seeds every year. Rather than concentrate their reproductive effort in certain years like the masting trees, they spread their seeds over more seasons but produce fewer each time.

## Masting in Cicadas?

Periodical cicadas (*Magicicada* spp.) could also be described as masting, for they emerge in enormous numbers to mate and lay eggs during certain years. Some species spend thirteen years underground as larvae, and others seventeen years. They emerge for a few short weeks during the thirteenth or seventeenth year in such huge numbers that the ground can be literally crawling with them. The stories about them in newspaper articles written during their emergence range in subject from people driven insane by their incessant buzzing to recipes for cicadas sautéed in butter. Different broods, or populations, emerge in different locations.

Since animals ranging from foxes and crows to humans eat cicadas, their masting, like that of the trees, may be an adaptation to reduce predation. The large and slow-moving adults emerge in such great numbers that enough eggs are laid to start the next generation even when many females are lost to predators. Alternatively, the cicadas may be avoiding specific pests that are unable to time their life cycles to the thirteen- or seventeen-year cycle of the cicadas. However, not all cicadas are the long-lived, masting species. In the Southeast in particular, the annual, or dog-day, cicadas (*Tibicen* spp.) emerge every year, using the same reproductive strategy as maples.

The buzzing call of cicadas is a means of communication between males and females. Only the males sing, and they make sounds by using muscles attached to a tympanum to cause a vibration. There are two of these drumheads located on the last segment of the thorax of the cicada's body. Most of the abdomen is hollow to create resonance and increase volume, just like the hollow body of a guitar or the kettle of a drum. When one male begins the chorus, others join in, and females follow the sounds to locate the singers.

FIGURE 1-7
An adult dog-day cicada (*Tibicen* sp.) emerging from its nymphal skin.

After mating, the females insert their eggs into the branches of trees. When the larvae hatch, they fall to the ground and use their enlarged front legs to burrow through the soil, where they feed on the sap of tree roots. Related to aphids, cicadas have a sucking tube for a mouth and use it to pierce plant tissues and suck up juice. (The mouth tube is not strong enough to pierce human skin.) During the summer months, the papery "shells," or exoskeletons, of the larvae can be found on the trunks of trees, where the larvae have metamorphosed into winged adults (Figure 1-7). The metamorphosis is rapid, with most nymphs climbing up the trunks during the night, splitting open their skins, and taking to the wing as adults early the next morning.

## Population Peaks in Voles as Long-term Cycles

Lemmings in Arctic regions are perhaps most famous for their population explosions and mass migrations, but related rodents such as the meadow vole and woodland or pine vole occur in the Appalachians and also undergo impressive population fluctuations. Voles (Figure 1-8) look something like mice but have short tails, small, furry ears, and rounder, bulkier bodies. Woodland voles (*Microtus pinetorum*) are brown, are about five inches long, and have a very short tail of only about an inch. These voles build extensive tunnel systems just below the leaf litter. Their enlarged front feet and small

FIGURE 1-8

A woodland, or pine, vole (*Microtus pinetorum*). Note the small ears, short tail, and bulky body, which distinguish it from a mouse. The very short tail and enlarged front claws identify the species.

eyes and ears enable this burrowing lifestyle. Voles live not just in woodlands but also in old fields that are densely vegetated.

The causes of rising and falling vole populations have not been determined, although several have been suggested. The most common explanation is usually that abundant food leads to an increase in voles, who then over-consume the food. When they run out of food, most starve and die. Voles eat a variety of plants and their seeds. In heavy vole years, I've seen them girdle small fruit trees by eating all the bark from the stem base and roots of the trees. Because they eat so many types of plants, general food supplies are unlikely to limit their numbers, but their favored foods are almost certainly reduced because so many voles are consuming them. It has been suggested that the voles' switch away from favored foods introduces a diet richer in toxins and poorer in nutrients, and this change may lower reproduction. Alternatively, voles' predators may control their population. It is clear that years of vole abundance support years of hawk, snake, feral cat, and fox abundance, and these populations may become large enough to reduce the number of voles. Another hypothesis suggests that as the population increases, aggression within the population escalates, thereby lowering reproductive success. Still another is that in large populations, disease is transferred more easily from one individual to another, and the stresses on the population may even weaken individuals physiologically, making them more susceptible to disease. Teasing out the single, most important cause of control has been difficult for ecologists and may be impossible if, as is likely, more than one controlling factor is at play.

## Geological Features as Extremely Long-term Cycles

The Appalachian Mountains were uplifted and eroded on at least four different occasions over millions of years. Mountain building and erosion therefore can be viewed as a long-term cycle that affects the biodiversity of the region. Certainly the old age and persistence of the mountains contribute to the high diversity of life that calls these mountains home (see Introduction).

The first cycle of Appalachian mountain building occurred nearly one billion years ago, during an event known as the Grenville Orogeny. A long mountain chain, estimated at 20,000 feet high, was pushed up as continents crashed together to form the supercontinent Rodinia. The fragmented remains of this Rodinian Appalachian Range can now be found from eastern North America to South America, around Antarctica and through southern Australia. A billion years ago, the only organisms on Earth were bacteria. Without plants to hold the soil, wind and water quickly eroded these mountains down to flatness. Rodinia also began breaking apart, drifting into smaller continents as the Iapetus Ocean opened up between them.

The second cycle of mountain building in the Appalachians, the Taconic Orogeny, occurred some 460 million years ago, when the first jawed fishes appeared in the oceans. Over the previous half billion years, Africa had separated from North America, but during this orogeny, it began a long process of drifting back and smashing into North America. There were still no plants or animals on land, and again these early Appalachians, which are estimated to have reached 15,000 feet in height, eroded to near flatness. All this sediment washed to the sea, creating a vast level plain on both sides of the former mountain range.

Between 420 and 370 million years ago, a third cycle of mountain building created a new chain of Appalachian Mountains. This Acadian Orogeny was driven by a collision between North America and fragments of the African continent that lay like islands between the two large continents. This episode is controversial among geologists, for whereas there is good evidence of the Acadian Orogeny in the Northern Appalachians, there is not substantial evidence for it in the Southern Appalachians. By this time, there were forests on land, and the first amphibians moved onto land among the conifers, giant clubmosses, and giant ferns that composed these early forests. The climate was warm and wet. For millions of years, primitive plants died and fell into the swampy waters that covered much of the region. It was these forests that formed the great coal deposits of the Central Appalachians, from Pennsylvania into West Virginia.

The Acadian Orogeny was a prelude to the major event that created the Appalachian mountains of today. This episode, the Alleghanian Orogeny, occurred around 330 million years ago, when the approaching continent of Africa finally slammed full bore into North America and formed the supercontinent of Pangea. That colossal collision pushed up peaks estimated to be 26,000 feet in high, as tall and craggy as the Himalayas are today.

Since then, for the last 330 million years, the Appalachians have been eroding down to their current maximum height of around 6,000 feet. (The highest peak in the east, Mt. Mitchell, is 6,684 feet high and occurs in the Southern Appalachians.) Although mountain building ended long ago in the Appalachians, things have not been quiet in geological terms. Obviously, Africa is no longer connected to North America as it was 330 million years ago. Instead, it has been moving away, opening up the Atlantic Ocean in its wake as Pangea broke up, just as the Iapetus Ocean was formed earlier by the break-up of Rodinia.

As climate cyclically warmed and cooled, ocean height rose and fell. About 90 million years ago, the oceans reached their all-time height when all the ice caps and glaciers melted. Ocean level was about 800 feet higher than it is today, meaning that the Atlantic washed the feet of the Appalachian Mountain chain. The Blue Ridge Escarpment, a sharp drop along the eastern front of the Southern Appalachians, where the mountains drop into the piedmont, may have been formed by the action of waves breaking against the base of the mountain chain, although geologists are debating this hypothesis. Should you find yourself at Caesar's Head State Park, located on the escarpment at the border of North and South Carolina, imagine that the vast piedmont visible below is awash in ocean water instead of clothed in greenery, and sea gulls rather than hawks are flying along the ridgeline.

Geologists anticipate that another mountain building event will begin in another 10–20 million years. The oceanic crust along the eastern edge of North America has been passive for more than 200 million years but should soon begin to sink back into the deeper mantle. "Soon" is a relative term, but geologically speaking, 10 million years is not far in the future. When that time arrives, the Appalachians will once again be home to volcanoes, earthquakes, and rising mountains, and the animals and plants that live here will adapt again to these changing conditions, migrate, or perish.

## The Nature of Cycles

Nature is rhythmic. The days spin from daylight to darkness and the years cycle through the four seasons as the environment swings over eons. Life responds and adapts to the changing but predictable conditions. Thoreau once said that if he were dropped any time into the forest of New England, he could determine the date within a range of two days. He heard the music as few others have, before or since. Yet the animals and plants of the region easily match Thoreau's ability and surely follow their environmental scores.

Because the events of the yearly cycle are strikingly broad, I use seasonality as the organizing theme of this book. Within the seasons, different organisms come to play at different times of day. Longer-term cycles affect the players by influencing biodiversity, but the long-term cycles are often the most difficult to visualize. Thus the seasons are the focus. What will you see when you step into the Southern Appalachians one fine spring morning?

# Cycles of Spring

## March, April, May

2

Spring transforms the Southern Appalachians from a boreal winter austerity to the vibrancy of a tropical forest. The stark gray hillsides of winter become a palate of green pastels, interspersed with luminous patches of contrasting flowering trees. The new leaves sparkle with raindrops, flowers open, and fiddleheads unfurl. As the tender green plants sprout from the brown forest floor, insects emerge to eat them, and then birds return to consume the insects while serenading us and each other. The abundance of food and warmth sparks the drive to reproduce, and the season throbs with youthful energy and new life.

### Pollination and Flower Form

While sneezing, washing cars, or cleaning windows, it is easy to forget that the effect of spring pollen on us is purely accidental. The whole point of prodigious pollen production is to transfer genes contained in the pollen to eggs held in the female part of the flower of a compatible plant. Pollen fertilizes the egg of a plant in much the same way that a sperm fertilizes the egg of an animal. Clouds of windblown pollen can be likened to the waterborne milt or spawn of many fishes, clams, and other aquatic animals. Wind-pollinated plants must produce huge quantities of pollen, for they rely on chance and the vagaries of the wind to deliver the pollen to the egg.

Insect-pollinated plants, by contrast, are much more selective. They produce less pollen but must attract and employ insects to transfer the pollen from one flower to another. Instead of spending their reproductive energy on copious amounts of pollen, they allocate some of that energy to producing colorful flowers and sweet nectar, which attract their animal pollinators. Primitive flowering plants, such as the magnolias, are pollinated by insects, which suggests that flowers evolved to attract insect pollinators.

How pollination is achieved is determined in part by the ancestry of the plant (all orchids, for instance, are insect pollinated and all pines are wind pollinated), but also by other factors, including the size of the plant. Large organisms, with substantial stores of energy, can afford to produce the huge amount of pollen necessary for wind pollination to be effective, and the large trees of the canopy often use the wind for pollination. By contrast, smaller plants typically produce fewer reproductive cells but ensure their fertilization by allocating relatively more energy into accessory structures, such as flowers and nectar.

Many canopy trees bloom early in the spring season because they are wind pollinated. Maple trees, for example, are notoriously early bloomers, usually beginning in March in the mountains and even earlier along the coastal plain. Spring winds are often strong enough to disperse pollen over great distances, resulting in beneficial mixing of genes from trees over a wide area. Since fully formed leaves would interfere with wind currents, wind-pollinated trees usually flower before the leaves are fully open. Their flowers are small and inconspicuous because they need not attract pollinators. In fact, petals and other flower parts would just get in the way of the windblown pollen. Thus, while most deciduous trees do not leaf out until May in the Southern Appalachians, many are active in other ways.

## Pollination in Serviceberry and Silverbell

Sarvis, or serviceberry (*Amelanchier arborea*), is a conspicuous blooming tree of early spring. Though small in stature, rarely reaching more than forty feet in height, its open, simple white flowers cover the tree and create an unmistakable white or slightly pinkish patch in the forest (Plate 1). If you're long-range botanizing from a car, it is easy to assume that the small white trees are dogwoods, but a quick up-close inspection reveals the distinctive flowers. Like other members of the rose family, serviceberry flowers have five small petals, are colorful, and have a pleasant odor. Because the flower is very open, exposing the stamens and stigma to the wind, and it blooms in early spring, it may be wind pollinated, but because most other members of the rose family are pollinated by insects and the serviceberry flower is both scented and colorful, insects likely play a role. Perhaps serviceberry just takes whatever pollen comes its way!

Serviceberry is primarily a tree of the Appalachians, but it occurs in all the eastern states. The common name comes from several sources. Service, or sarvis, depending on your dialect, blooms about the time that the ground thaws out enough to be worked. Historically, people who died during winter were often held for burial until spring. Wandering preachers provided the funeral services all over the Appalachians during the time that serviceberry blooms. In addition, as roads became more passable in springtime, family and friends could attend memorial services more easily than during the winter season. Serviceberry is also called shadbush because the blooming period coincides with the run of shad upstream to spawn. The red berries of serviceberry, which appear in June, provide a valuable service of food long before most other berries are in season and accounts for another name for serviceberry: Juneberry.

One of the most beautiful trees in Appalachian cove forests is silverbell (*Halesia tetraptera*), named for the pendent white bells that hang from its branches in bunches (Plate 2). The bells are about an inch long and are composed of four white petals that are fused together around a cluster of stamens, laden with yellow pollen, and a pink central stigma. When they fall, they cover the ground like snow, making it seem more like Christmastime than springtime. The inch-long, strangely shaped fruits appear in summer, and the scientific name refers to the four (tetra) wings (ptera) of these fruits.

Silverbells are usually understory trees, rarely reaching the canopy. They can grow to about ninety feet but are usually much smaller. The twigs and small branches of silverbell have distinctive white streaks on them, and the bark of larger trees is blocky.

## Wind Pollination in Maples, Oaks, and Grasses

Red maple (*Acer rubrum*) is a glorious, colorful tree with red flowers in spring and red leaves in fall. The small, reddish flower buds begin to swell in late winter and open by March in most Appalachian locations. Each cluster of small flowers is either male or female (Figure 2-1). Although both can occur on the same tree, I have often observed that most trees produce either male or female flowers rather than both. When the entire tree is in bloom, it provides a blush of red on the face of an otherwise gray forest. For me, it isn't the clusters of flowers on the trees themselves that draw my attention but rather those strewn along forest paths after having been nipped off by feeding squirrels and finches.

FIGURE 2-1

Male and female flowers of a red maple (*Acer rubrum*) occur in clusters either on the same tree or on different trees. Male flowers (left) have pollen-bearing stamens, and female flowers (right) have a Y-shaped stigma.

Red maple takes advantage of wind not just for pollination but also for fruit dispersal. The winged fruit, or samara, appears in late spring, just as the leaves of the tree are enlarging. Pairs of samaras hang down from a two-inch-long stalk. When the seeds are ripe, a strong wind will tear them free and set them spinning like the blades of a helicopter. As they slowly spiral down, wind carries them farther from the parent tree, dispersing the seeds (and resultant seedlings) into new areas. Seeds of some maples have reportedly been transported over twenty-five miles from their parent tree, although distances are typically much less.

Red maple grows nearly everywhere in eastern North America. It tolerates a greater variety of habitats and exhibits a wider distribution than any other tree in the region. It can grow with its roots submerged in water, and is happy in the coastal swamps and flood plains of the southern states. It is equally at home in the coves of high mountains, as long as it has sufficient water. The trunks of maples are usually smooth and light in color, though they are often covered with greenish lichens. Red maples can reach about a hundred feet in height, but most mature trees are only half that size.

In addition to the characteristic samaras, maples have palmately lobed leaves, with the lobes arranged like short but fat fingers jutting from the palm of a hand. Because there are so many species and the particular pattern of lobes differs in each, a field guide is useful to distinguish them. In the most extreme example, the "lobes" of eastern boxelder (*A. negundo*) are independent leaflets and the leaves are compound (that is, each leaf is made up of several distinct leaflets), but the leaflets are lobed, too.

The most familiar maple of all is sugar maple (*A. saccharum*), known both as the source of maple syrup and of highly desirable lumber. "Curly maple," "rock maple," and "hard maple" usually refer to wood of the sugar maple. Striped maple (*A. pensylvanicum*) is a common understory tree of the Appalachians with greenish, white-striped bark and prominent, drooping flowers and fruits. Mountain maple (*A. spicatum*) is an uncommon small tree found in higher-elevation forests of the Appalachians. Silver maple (*A. saccharinum*), whose silvery leaves dance in the winds of summer and fall, is widespread but has been introduced into the Appalachians from farther west and is more common in urban and suburban settings than in the forest.

Oaks are also wind-pollinated trees, but they bloom later than red maple, while their leaves are emerging. Their tiny, inconspicuous female flowers are usually borne in pairs along the new growth on which the leaves are arising. The male flowers hang down in dangling yellowish green catkins (Plate 3). The diversity of oaks is impressive, making identification of species the most difficult among our common trees. To further complicate the issue, oaks often hybridize, forming intermediates between species. Their leaves and acorns are described in Chapter 4.

Grasses, sedges, and rushes, which are also wind pollinated, are similar to large trees in that they normally grow in open situations where wind has easy access to their pollen. Rarely do grasses grow in mature forests; instead, they grow in open fields where they are often the tallest plants. Similarly, their flowers are reduced to just the male and female parts; there are no sepals or petals to attract insect pollinators or interfere with the wind.

While grasses, sedges, and rushes look superficially similar, they have distinct differences. Sedges are usually triangular in cross section. Rushes, like grasses, have round cross sections, but they do not have joints in the stems. Grasses, such as bamboo, corn, and fescue, have distinct joints in the stem from which the leaves arise, and the stem is hollow. A good way to remember the difference is to use this verse: "sedges have edges, rushes are round, grasses have joints all the way to the ground." An alternative, "grasses are hollow down to the ground," also works!

## Insect Pollination in Tulip Trees, Magnolias, and Flame Azaleas

Tulip trees (*Liriodendron tulipifera*) are one of the largest, most common, insect-pollinated trees in the Southern Appalachians. They produce beautiful tuliplike flowers. Also called yellow poplar, I prefer the name tulip tree because it reminds me that they are not really poplars but belong to the magnolia family. Foresters, however, are loath to call it tulip tree because there is a tropical hardwood already known by that name in the trade.

Tulip trees leaf out in early May, and the beautiful flowers follow in late May or even June. Each large flower is about three inches wide and consists of six petals that overlap to form a cup. Yellowish for most of their length, the petals have a strip of orange that aligns nearly perfectly with the pollen-bearing stamens (Plate 4), probably to help attract pollinators. Insects buzz busily in the large blooms as they gather the nutritious pollen and nectar. The nectar is produced on the surface of each petal from its base up to the orange strip. Each flower contains multiple female stigmas on the central cone that will eventually become inch-long woody seeds. The winged seeds, like those of maples, are distributed by winds of late summer and fall.

Tulip trees are common in the coves of the Appalachian forest and grow best where the soil is fertile and moist. They grow rapidly, and their soft wood is more like pine than oak. They are the tallest Appalachian trees, reaching about two hundred feet in height and ten feet in diameter when allowed to grow for centuries. Joyce Kilmer Memorial Forest, in western North Carolina, has the most impressive tulip trees I have ever seen, and the largest tulip tree yet identified grows in the Great Smoky Mountains National Park off the Boogerman Loop Trail.

Fraser magnolia (*Magnolia fraseri*) is the most common Appalachian magnolia, although several species occur in the Appalachians and in adjoining regions. The native Appalachian species are all deciduous, but the evergreen bull-bay, or southern magnolia (*M. grandiflora*), is the one most frequently planted horticulturally outside its native range and does well in the lower elevations of the Southern Appalachians.

Fraser magnolia is identified by the small lobes on each side of the leaf petiole on the large, foot-long, otherwise unlobed leaves; it is the only common Appalachian magnolia with these lobes. The large, simple leaves make it a dramatic plant in the landscape. In the fall, the leaves turn such a pale yellow that sunlight seems to emanate directly from them and set them aglow. The large leaves are adapted to absorb filtered light, and the smooth-barked trees rarely form part of the canopy. In the understory, those large leaves are protected from wind damage. They are a late-successional species and appear after the forest has otherwise matured.

Like tulip trees, Fraser magnolias bloom in late spring after the leaves have fully opened. Creamy-white petals, each about four inches long, surround the stamens and pistils. The fragrant flowers attract their insect pollinators and human admirers alike. The seeds develop inside a cone-shaped, four-inch-long fruit, which exudes a spicy fragrance. Once ripe, the brown fruit dries and opens to release the bright red or pink seeds, which dangle for a time from a long thread. These seeds are high in fat, making them attractive to migrating birds (see chapter 4).

Another beautiful specialty of the Southern Appalachians is the shrub flame azalea (*Rhododendron calendulaceum*). Its colorful flowers range from pale yellow to nearly red, although various shades of orange are most common. Spring hillsides burst into color when these shrubs bloom, setting the forest aflame. Each flower has several long stamens and one stigma that stick out beyond the petals. I have observed ruby-throated hummingbirds, bumblebees, and swallowtail butterflies visiting the flowers, but this is an uncommon occurrence. Most small insects avoid them altogether. While the flowers are showy, they do not produce a large quantity of nectar or pollen, which may explain the rarity of pollinators and mature fruits. In addition to flame azalea, several other species of native azaleas occur in the Southern Appalachians, and their blooms are equally attractive. Our native azaleas are deciduous species of *Rhododendron*. The evergreen rhododendrons are described in chapters 3 and 5.

## Of Peas and Pollinators: Locusts and Other Legumes

Black locust (*Robinia pseudoacacia*) is a common insect-pollinated tree that blooms in late spring. It belongs to a large group of plants, the legume, or pea, family. Redbud trees, clover, vetch, wisteria, and kudzu are also legumes. Their flowers all have the same basic structure, with a lower lip tucked into an upper petal shaped like a hood. The pollinator forces its way under the hood, brushing against the pollen-bearing anthers and pollen-accepting stigma before it can reach the nectar source. Only pollinators that fit the flower's shape are able to pollinate it and receive the nectar reward. The shape of the bilaterally symmetric flower reflects that of its bilaterally symmetric pollinator. Bees, including non-native honeybees, native bumblebees, and several small native bees, are the chief pollinators of legumes, attracted to the bright colors and sweet nectar of the flowers. Perhaps it is hard to believe, but the floral perfume of kudzu, an invasive non-native legume, is heavenly!

It may seem surprising that a large tree such as black locust, a vine such as kudzu, and small nonwoody plants such as white clover are all classified as peas, members of the same family, but the families of plants, which are made up of closely related species, are distinguished by the structure of their flowers and other features, not on the form or size of the entire plant. Botanists use flower, fruit, seed, and sometimes leaf structure to determine relationships between plants because those structures vary less than the structure of stems. For example, in addition to the similar shape of pea flowers, the seeds of most peas are born in a pod, which is what we eat from the domesticated species, and the leaves of most peas are compound. Molecular-based studies are becoming increasingly important in determining plant relationships and have led to new classifications of some plant species, genera, and even families.

Legumes have another characteristic that contributes to their success and great diversity (and usefulness to humans). They have developed a symbiotic relationship, a mutualism, with bacteria that benefits both the plant and the bacterial partner. All pea plants have special nodules on their roots in which *Rhizobium* bacteria grow (Figure 2-2). Each plant species hosts a different species of *Rhizobium*. The nodules protect the bacteria from oxygen, which damages their biochemical machinery. The bacteria also benefit from energy (in biological units, called ATP) and nutrients (mostly carbohydrate sugars) provided by the plant. For all this hospitality, the bacteria only contribute nitrogen in return, but it is nitrogen the plants crave. Nitrogen is the nutrient most in demand by plants because it is essential for the manufacture of proteins and DNA, yet it is in short supply in most soils. It is so important for plant growth that it is the basis for commercial fertilizers, and is listed first in the percentage breakdown of components. For instance, 10-10-10 fertilizer has 10 percent nitrogen, 10 percent phosphorus, and 10 percent potassium; the rest is filler.

The fact that nitrogen is in short supply for plants is really rather surprising, since our atmosphere is about 80 percent nitrogen. Atmospheric nitrogen ($N_2$), however, is in an extremely stable form in which two atoms of nitrogen are stuck tightly together to form a molecule. In that form, it is unusable by plants or animals. Most organisms are unable to break this strong molecule apart and free each N atom for their own use. There are only two natural means to break down $N_2$. The enormous energy in a lightning strike is enough to blast $N_2$ apart and recombine it with oxygen. The

FIGURE 2-2

All peas, including white clover (*Trifolium repens*), pictured here, have special nodules on their roots that house nitrogen-fixing *Rhizobium* bacteria.

other is nitrogen-fixing bacteria, which convert $N_2$ into the usable forms of ammonium ($NH_4^+$) and nitrate ($NO_3^-$) ions. Some nitrogen-fixing bacteria live freely in the soil, but those that have formed a mutually beneficial partnership with plants are the most important in terms of how much nitrogen they can provide.

The success of the legume "strategy" of forming an alliance with *Rhizobium* to assure its supply of nitrogen is paralleled in few other plants. Alders and wax myrtles form a similar partnership with the bacterium *Frankia alni*. These bacteria provide nitrogen in exchange for a protective root nodule in which to grow, as well as a few nutrients provided by the host plant. In return, the bacteria provide nitrogen for use by the plant. The leaves of alder, like those of most legumes, are often so full of nitrogen that they actually enrich the soil of the forest with nitrogen each autumn when they fall to earth and decompose.

Black locust, the most common locust species in the Appalachians, is a good example of the advantages a legume has over other plants that do not fix their own nitrogen. The fragrant, white flowers are born in clusters on the branches. As a pioneering species, it is a common early arrival on newly cleared land, such as road cuts. It can grow in poor soil because it, like other legumes, provides its own fertilizer. It grows rapidly, quickly outcompeting other plants that find fewer nutrients in the soil, and easily reaches ninety feet in height. Small trees and limbs often bear pairs of sharp spines, protecting them from browsing animals that find the foliage nutritious. The trees are commonly used as fence posts because their wood is hard and not prone to rot or insect damage. Removing a locust stump is hard work!

Other animals besides grazing ones take advantage of the fertilizer that is stored in the leaves of black locust. Locust leaf miners (*Odontota dorsalis*) are small orange and black beetles about 1/4 inch long that damage leaves of black locust. The larvae, which tunnel between upper and lower leaf surfaces, do most of the damage. They usually begin along the central midrib, but then extend their feeding activities over the whole leaf. Often, the leaves die because most of the photosynthetic tissue in between the upper and lower epidermis has been consumed. In some years, the number of beetles is so large that almost every leaf on the tree is damaged. These damaged trees, with their dead, bronze leaves, are visible late in the summer after the beetles have had plenty of time to eat. Miners prefer trees that fix higher quantities of nitrogen.

## Ephemeral Wildflowers: Ramps and Trout Lilies

Many spring wildflowers bloom, set seed, and wither away by the time the trees are fully leafed out, rarely lasting above ground past the first days of June. Because both their flowers and their leaves are present for a short period of time in the spring, they are called spring ephemerals. Many mountain wildflowers are spring ephemerals, including ramps, trout lily, spring beauty, hepatica, wood anemone, Dutchman's breeches, squirrel corn, bleeding heart, and windflower. It is important to note that in true spring ephemerals, leaves are only present for a very short period of the year.

Spring ephemerals are small plants that grow near the ground in mature deciduous forests and are long-lived, blooming year after year. They absorb sunlight during the few spring months between winter and summer, before the large trees take it over. By summer, rather than wasting energy to maintain leaves that no longer receive sufficient light, spring ephemerals store the energy they have made in thick roots or bulbs and wait for the next spring. For example, squirrel corn is named for the yellow bulbs on its roots that resemble corn kernels and are loaded with nutrients. Most spring ephemerals also rely on ants to disperse their seeds, and a complex relationship has evolved between these two organisms.

Most people dig up the bulbs and discard the leaves when they harvest ramps to eat, but if the bulbs are eaten, the plant is destroyed. Because of overcollecting, many collection locations have been closed to harvesting and most others require a permit for collection. As with any other life-form, harvesting ramps faster than they can reproduce leads to their eventual extinction. I recommend a sustainable, less destructive method of harvesting that provides equal gustatory delight. From my own patch of ramps, I collect only one of the two leaves of each plant. (The leaves are delicious. I honestly prefer them to the strongly flavored bulb.) By leaving the bulb intact and occasionally removing competitors, I avoid killing the plants and actually maintain and encourage a permanent source. Leaving one leaf allows the plant enough energy to set seeds, perpetuating my supply. You could say that the ramps patch and I have evolved a happy mutualism.

If you are curious about ramps, check out the annual festival in Waynesville, North Carolina, which is typically held the first weekend in May. Sylva and Robbinsville hold festivals on the last weekend of April, and other communities often sponsor suppers at which ramps are featured. Ramps festivals occur in many mountain communities of Tennessee, Virginia, and West Virginia as well. If you go to one, be sure to ask whether the ramps served at the festival were sustainably harvested!

A springtime delicacy of the Southern Appalachians is ramps (*Allium tricoccum*), a close relative of garlic and onions. They are called ramps because they appear during the zodiac sign of Aries the Ram, in late March and early April. "Ram's son" was shortened to "ramps." In the spring, my neighbors and many other local mountaineers make pilgrimages into nearby forests in search of the pungent bulbs and leaves. Ramps grow in rich forests, where the topsoil is deep and fertile. Although most abundant in the Southern Appalachians, they occur as far north as Canada and as far west as Minnesota.

Ramps and their relatives have been used medicinally to treat a host of ailments, and one of these old-time remedies has crossed over into the world of modern medicine. One of the most common uses of both garlic and ramps was to expel internal worms, and a concentrated form is now produced commercially. It is called allicin, from the scientific name *Allium*, the group name for all onions, garlic, and ramps.

Ramps are most easily identified by their leaves. A bulb typically produces two broad, flat leaves, which are a soft, silvery green, easily bruised, and measure from 1 to 2 1/2 inches wide and from 5 to 10 inches long. The leaves of ramps look very different from those of field garlic (*A. vineale*), which grows so profusely in suburban lawns all over the Southeast. The leaves of field garlic are skinny, hollow cylinders, usually 1/8 inch in width and up to 12 inches long. As spring ephemerals, the leaves of ramps are present only briefly, withering away and disappearing in late May. A cluster of small, white flowers is produced after the leaves shrivel.

There are many lilies and related plants emerging in the forest at this time of year, some of which are poisonous to consume. Probably the plant most easily confused with ramps is false hellebore (*Veratrum viride*). The leaves of both it and fly poison (*Amianthium muscitoxicum*) appear in early spring and persist through most of the summer season. In early spring, the leaves look green, tender, and delicious, but don't be fooled! Both plants are extremely toxic, with a few reports of human and livestock death from eating the plants. The roots of false hellebore have even been used to produce an insecticide. The two plants' leaves look somewhat similar when they first appear, but as they unfurl, they become more distinctive. The leaves of false hellebore are oval and up to a foot long and over four inches wide. The leaves of fly poison are about a foot or more long but are less than an inch wide. The blooms make it easy to identify the plants. Both send up tall flower stalks, but false hellebore's flowers are green and spread out along the tall stalk and fly poison's flowers are white and clustered into a compact bunch at the tip of the flower stalk.

Trout lilies (*Erythronium umbilicatum*) are also ephemeral wildflowers. While they occur in mature forests, I have found them more commonly in meadows and floodplains of streams. They emerge in early spring before the grasses and tall perennials overtop them. By the time summer arrives and the meadow is waist-high in vegetation, the trout lilies have bloomed, set seed, and retreated back into bulbs to wait for the next spring. Each elongated, mottled leaf does, indeed, resemble the speckled back of a brook trout in shape and color. The flowers, though, are nothing fishlike. Instead, they look like shooting stars. Bright yellow, the six petals bend back away from the exposed central stamens and pistil. Held above the ground on a short stalk, they nod in gentle breezes. When the plants are blooming, two leaves are usually present, but in years that the plant does not bloom, only a single leaf is produced. The pair of leaves may protect the emerging flower as the stalk pushes up through abrasive soil, and two leaves also double the input of sunlight to the plant, replacing some of the energy used to produce the flower and seed.

## Other Wildflowers: More Lilies, Trilliums, and Jack-in-the-Pulpits

In addition to the spring ephemerals that quickly fade away, many other spring wildflowers persist longer. While their flowers may be short-lived, their leaves remain present throughout the spring and summer months. Sometimes these flowers are grouped with spring ephemerals, but the plants are not truly ephemeral because they remain active for most of the growing season.

Many of the most beloved spring wildflowers are monocots, one of the two major divisions of flowering plants. Lilies, trilliums, ramps, orchids, and grasses are monocots. The leaves of monocots have parallel veins that run in orderly rows, and their flowers are composed of three petals, three sepals, six anthers, and a single pistil in three parts; in the case of lilies, the petals and sepals are indistinguishable from each other and are called tepals.

Speckled wood lilies (*Clintonia umbellulata*) begin blooming in mid-May at low elevations in the Southern Appalachians. The six or fewer shiny, green, broad leaves are edged with delicate silvery hairs. In the center of the basal rosette of leaves, a flower stalk about six inches high supports a head of several small white flowers. At higher elevations and farther north, Clinton's lily or bluebead lily (*C. borealis*) is common. Like many organisms of northern climes, this lily is bigger than its southern relative. The flower is about 1/2 inch long, droops downward, and is yellow rather than white.

Trilliums have three large leaves and a single large flower with three colorful petals, which lend them their name. There are numerous species, but a few deserve special mention. Among the earliest trilliums to bloom are little sweet Betsy (*Trillium cuneatum*) and yellow toadshade (*T. luteum*). Both are sessile-flowered trilliums: the maroon or yellow flower sits right atop the leaves, which are dark green and mottled with pale splotches of green, and both are restricted to the Southern Appalachians. Wake-robin (*T. erectum*) and great white trillium (*T. grandiflorum*) are common and more widely distributed, occurring from the southern mountains up into Canada. The flowers (pink or white on the great white trillium; usually maroon or white on the wake-robin) project up on a stalk from the leaves and are hard to miss in a walk through the woods or even a slow drive along mountain roads in spring (Plate 5). The pretty red and white flowers of painted trillium (*T. undulatum*) are also borne on upright stalks, but this trillium prefers the acid soils under rhododendrons and can be harder to see unless you are really looking for them. It is widely distributed, occurring as far as Canada, and is the most acid-loving of all the trilliums. The largest-flowered trillium is late-blooming Vasey's trillium (*T. vaseyi*), another species restricted to the

Southern Appalachians. It shyly nods its huge burgundy (or rarely white) flowers toward the ground as though it can hardly support their weight.

The fruit of Vasey's trillium is also huge, nearly an inch in diameter. Trilliums, like the spring ephemerals, depend on ants to disperse their seeds and have modified their seeds to attract them. The fleshy fruits, however, also attract mammals, birds, and even yellow jackets as dispersers. Painted trillium fruits, for example, are a brilliant red when ripe and attract the attention of a visually oriented animal. Trilliums, therefore, put their "eggs" (seeds) in more than one basket by attracting more than one kind of animal to disperse them.

Nonflowering specimens of jack-in-the-pulpit (*Arisaema triphyllum*) are sometimes confused with trilliums because their leaves are also three-parted (Plate 6), but jacks typically have a pair of three-parted leaves instead of a single stem and leaf. In deeply shaded forests, the leaves are only six inches or so high, but I've also seen them in Florida swamps where the leaves were nearly two feet high!

The flower, on a separate stem, makes jack-in-the-pulpit easy to recognize. The central "jack" is the flowering part of the plant and has multiple tiny flowers. The "jack" is surrounded by a vaselike spathe with a flap that arches over the top, similar in shape to the pitcherplant leaf. The spathe often has purplish brown stripes and dries up as the seeds ripen so that they are exposed when fully ripe. Its fancied resemblance to a medieval hooded pulpit is the basis for the common name.

The seeds develop inside bright red fruits that are held on the upright stalk. Only the female flowers develop into the nutritionally expensive fruits, and the number of female flowers that appear on a plant depends on its nutritional state. Young plants have fewer nutritional reserves available and usually produce only male flowers, but as the plants grow older and larger, and more nutritional reserves are available, individual plants switch gender to female.

## "Leaves in Three, Let It Be!"

Often, visitors from other areas enter the forests of the Southern Appalachians and panic when they notice all the trilliums, jack-in-the-pulpit, blackberry, seedling hickories, and Virginia creeper vines that have a superficial resemblance to the leaves of poison ivy (*Toxicodendron radicans*). Poison ivy is a woody vine that climbs trees or trails along the ground and is most common along the sunny edges of fields and forests. The three-parted leaves

FIGURE 2-3

The leaves of poison ivy (*Toxicodendron radicans*), which are three-parted and have pointed lobes, are produced alternately from a hairy, woody stem.

alternate along the stem and usually have pointed lobes (Figure 2-3). Unlike grapevines, which are also common in these habitats, the stems of poison ivy are covered in tiny roots, helping them adhere to tree trunks.

The poison in poison ivy, called urushiol, causes blistering and itching on human skin; the severity of the reaction depends on the dose received and the individual's response to it. Even the stems and roots contain the poison, and smoke, clothing, or pets that come in contact with the plant and then a person can cause outbreaks. Removing the oil by washing with soap and water, swabbing exposed skin with alcohol, or using jewelweed sap (see chapter 3) are all effective methods of preventing the dermatitis.

Along the coastal plain, where I grew up, a shrubby version of poison ivy known as poison oak occurs; there is still debate over whether it is another species or just a different growth form. Poison sumac is rare in the mountains, where it occurs in bogs. Most sumacs (see chapter 4) are not poisonous.

## Oconee Bells: A Charismatic Appalachian Wildflower

The Southern Appalachians are home to one of the rarest of American wildflowers, the evergreen Oconee bells (*Shortia galacifolia*), which forms a patchy groundcover. Although it can be locally abundant, has been widely trans-planted, and is not a federally endangered species, its natural distribution is extremely localized. It is native to only seven counties in North Carolina, South Carolina, and Georgia and is named after Oconee County, South Carolina. Not only is this pretty wildflower rare, but it is one of the earliest native plants to bloom, a longed-for indicator of spring.

Oconee bells was discovered by one of the most prolific botanists who combed the Appalachians in search of new plants. André Michaux, a Frenchman who settled in Charleston, South Carolina, first found the plant in June of 1787 in the mountains of the Carolinas. Two other American bota-nists, Asa Gray and John Torrey, named the plant for Kentucky botanist Charles Short after Gray found a single, flowerless specimen in Michaux's collection in Paris. They knew that no one since Michaux had seen the plant, and realizing its significance, they began to search for a flowering specimen. Collecting expeditions all ended in vain, and the elusive plant, like the Holy Grail, began to take on mythical proportions. Nearly one hundred years later, in 1877, a seventeen-year-old boy named George Hyams rediscovered the plant in McDowell County, North Carolina. Since then, botanists and historians have carefully retraced Michaux's route, following the descrip-tions in his journals, leading them to the original source of the plant along the upper Keowee River, near the old Cherokee town of Jocassee, in Oconee County, South Carolina.

The story of the legendary plant abounds with ironies. The first is that Charles Short, the frontier botanist after whom the plant is named, never saw the plant. He died fourteen years before it was rediscovered. Second, since Michaux, who first collected it, did not name it or collect a flower, American scientists named an American plant based on a French collec-tion. Third, a young boy, not a botanist, found the first flowering specimen. Fourth, Michaux's original collection location is now submerged by Lake Jocassee, extinguishing the discovery site and the population. Fifth, the col-lection date and location were not resolved until 1983!

Oconee bells (Plate 7) is closely related to galax (*Galax urceolata*), and the leaves of the plants look similar: evergreen, glossy, low-growing, about two inches in diameter, and rounded. The leaves of Oconee bells, however, are shinier and more obviously veined. The scientific name of Oconee bells, *S. galacifolia*, refers to this similarity of leaves; *folia* means leaf. Collection of galax leaves for the florist industry has recently become so prevalent that collection permits and a restricted harvest season have been imposed on National Forest lands in North Carolina.

Unlike galax's tall, slender stalk of tiny white flowers that resembles a pipe cleaner, the inch-long flowers of *Shortia* are borne singly on salmon-colored stalks that project about five inches above the ground. The bell-shaped flower has five waxy, pinkish white, fringed petals surrounding five yellow, shield-shaped anthers. A pink stigma projects outward from the center of the flower.

## Bloodroot: A Native Poppy

Another of the early wildflowers is bloodroot (*Sanguinaria canadensis*), a native poppy. A single leaf first emerges from the forest floor in a tight roll, like a green cigar, then it unrolls into a large, flat leaf measuring several inches across (Plate 8). Because they stand only a few inches above the forest floor, it helps to shift your perspective from eye-level to ankle-level when hiking in the spring.

The delicate flowers, which look somewhat like daisies with long petals, soon follow the leaves. The centers are yellow, and the eight or more, inch-long petals are white. A few gentle raindrops are enough to dislodge the petals, and after a heavy spring rain, the fallen petals lie like a pile of bones underneath the solitary leaf. The seeds are produced in a capsule on the end of the flower stalk after the flower withers, and bloodroot, like many other spring wildflowers, depends on ants to disperse its seeds. Like trout lilies and Oconee bells, bloodroot also reproduces asexually from outgrowths on underground stems.

If the rootlike stem of bloodroot is injured, it releases red juice. Because the root appeared to bleed, it was used in folk medicine to treat a variety of diseases. Its scientific name comes from the Latin term for blood (*sanguis*), as does the English adjective "sanguine" and the Spanish "sangria." The name *canadensis* refers to its distribution from Canada through the Appalachians. This red pigment is antibacterial, and bloodroot has been used commercially in toothpaste and livestock feed.

## Plants and Ants: Beneficial Relationships

Most of the spring ephemerals mentioned previously, as well as a number of other spring wildflowers, have developed a beneficial symbiosis with ants, exchanging seed dispersal for a food reward. The hard and slippery seeds of ant-plants have a tender fleshy extension on their sides that resembles and functions like the handle on a teacup (Figure 2-4). This handle, called an elaiosome, allows the ants to more easily carry the seed to their nest, where

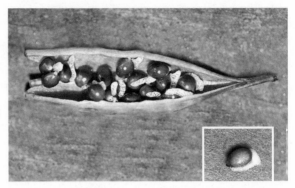

FIGURE 2-4

Several plants produce seeds with elaiosomes. As shown in this opened seed capsule of bloodroot (*Sanguinaria canadensis*), each dark seed also contains a soft, white elaiosome. The hard seed of vernal iris (*Iris verna*) (inset) also has a fleshy elaiosome.

they eat the tender and nutritious elaiosome. The rest of the seed is too hard to consume and is discarded in the ants' garbage pile. The elaiosomes are a predictable and rich source of fats and protein in spring and early summer before many other foods are available and are especially important for the ants' larvae.

The plants benefit, too. The ant hauls away the seed from the parent plant, where it is less likely to compete successfully for light and nutrients, and "plants" it in a pile of fertile compost when it is thrown out with the rest of the ants' waste. Seed predators have less access to the seeds because they are removed from the plants as soon as they ripen and taken to an ant nest, where few seed predators such as rodents will bother them.

The dwarf irises (*Iris verna* and *I. cristata*) are good examples of plants that rely on ants to disperse seeds. They are not spring ephemerals, for, like other irises, their leaves persist year-round. In locations where there is little competition for sunlight, the swordlike clusters of leaves pierce through the soil in early spring. I find *I. verna* most frequently on the dry banks of road cuts, where their intensely purple blooms give them away for a few days each spring. *I. cristata* prefers moister sites. Although each purple flower lasts only a day, the plant produces multiple flowers, providing a show that lasts for a week or more. The flowers are pollinated by flying insects such as bees, but the seeds are produced in a capsule near the ground, making them accessible to ants.

Most of our violets (*Viola* spp.) also produce capsules near the ground so that ants can easily disperse their seeds. Like other ant-plants, their seeds bear elaiosomes. Our numerous species of violets range in flower color from yellow to white to purple. Most bloom in spring, but their leaves persist through the summer months. The leaves, which arise from a distinct stem or

Currently, red imported fire ants in North Carolina and Virginia are restricted to the coastal plain and piedmont, but they are reported from every county, including the mountain counties, of South Carolina, Georgia, and Alabama. Tennessee's southern counties, including the mountain counties, also have fire ants. These ants thrive in open locations such as fields and roadsides, where their nests are warmed by sun. As long as the forest remains intact, they rarely invade it, but as roads and logging operations open up the forest, conditions improve for this exotic ant. In addition, since it is only the cold winters that currently prevent their spread, global warming may increase the northern extent of their range. As an example, Transylvania County, North Carolina, occasionally has small outbreaks of these dangerous pests along its southern border, but a cold winter eliminates them. In the hot and dry summer of 2007, colonies popped up in several locations. If this introduced pest successfully establishes itself in the heart of our mountains, it is likely that our gorgeous, ant-dependent plants will suffer.

from a common base, may be round, pointed, or deeply dissected. Leaf shape and position and flower color are used to identify the different species, and hybrids may occur.

While many species of plants rely on ants for seed dispersal, few depend on them for pollination. Among those that do are heartleaf and wild ginger, both of which produce inconspicuous flowers that attract ants, beetles, and flies. Unlike the colorful, showy flowers of other species sticking up high in the air to attract flying insects, the pretty little brown jugs of these plants lie on the ground, tucked in among the heart-shaped leaves. They are visible only if you know to scoop away the collected leaves from the base of the plant. They never fail to delight me as a treasure when I find them. We are especially lucky to have the rare French Broad heartleaf (*Hexastylis rhombiformis*) in my neck of the woods.

The introduction of red imported fire ants (*Solenopsis invicta*) from South America, however, threatens this important relationship between plants and ants. Fire ants outcompete and eliminate most native species of ants they encounter. The refuse piles of these ants do not promote germination and growth of seedlings. Finally, when fire ants collect seeds with elaiosomes, they usually destroy the entire seed. In this relationship, only the fire ants benefit.

## The Orchids: Wildflowers that Trick Insects and Parasitize Fungi

While many different wildflowers have developed a mutually beneficial relationship with ants, providing food in return for seed dispersal, other plants are not such honest partners. Some flowers trick insects into working for them without paying anything in return for the service. Perhaps surprisingly, some of our most beautiful and highly prized native flowers are just such devious mistresses.

Seemingly overnight, globes of pink or yellow atop slender stems erupt from the forest floor as pink moccasin flower (*Cypripedium acaule*) and yellow lady's slipper (*C. parviflorum*) nod their heads to the advances of bumblebees, acquiescing to their insistent intrusions. A hovering bee bumps into a flower's hollow slipper, or moccasin, slips through the open cleft of the pouchlike lip, and is trapped inside. Unable to exit by its entry route, the bee climbs upward toward the flower's hood, fumbling through the tight furrow formed by the fusion of flower parts, first passing the female stigma and then brushing against the male pollen sacs before finding the escape opening at the top of the flower (Plate 9). This tight embrace ensures the bee unwittingly picks up orchid pollen. Entranced by another pink slipper, the bumblebee repeats its performance, this time depositing its pollen baggage on the new flower's stigma, thus fertilizing the flower and allowing it to set seed. Again, the freed bee seeks another flower and the cycle is repeated.

These lady's slipper orchids are elaborately designed to use the bee for their reproduction, but the bee receives nothing in return. Coldly unsympathetic, these alluring flowers rely on the naïveté of the bee, enticing it time and again into delivering the flower's future progeny without so much as a drop of nectar for a reward. But bees are not fooled forever and eventually stop visiting these orchids since they receive no reward of nectar. Such an arrangement results in relatively few flowers being pollinated (as evidenced by the low number of seed pods), but, because each seed pod produces an enormous number of tiny seeds, the plants remain relatively common and widespread.

One reason there are so many different species of orchids (there are more kinds of orchids than there are of any other family of plants) may be that they tie their reproductive success to one or, in some cases, a few species of insects. Because insects are the most diverse land animals, orchid pollination tied to only one insect species reproductively isolates each species of orchid and maps orchid diversity onto the impressive diversity of insects. Another explanation for the striking diversity of orchids is their habit, especially in

PLEASE DO NOT DIG ORCHIDS

It is best to leave our native orchids where you find them. Because they are such beautiful flowers, it is natural to want to possess them, but it is important to resist that urge! Our native orchids shouldn't be dug up for two reasons: First, it is nearly impossible to move them successfully because of their fungal partners. Digging up a plant and just enough soil to fill a pot is the surest way to kill it because the fungus will be disrupted. Second, many of these orchid species are rare and restricted to a few locations. Moving an orchid from one of these locations not only reduces a small population significantly, but it also removes that individual from the gene pool of the species, preventing it from reproducing more of its kind. Every individual that is eliminated lowers the genetic diversity of the species and its chance for continued survival. Removing orchids that are growing on public lands such as state or federal forests is illegal. Luckily, the hardier species of native orchids have been successfully cultivated and are now offered for sale. Check the credentials of the company from which you purchase orchids, though, to ensure the plants are not collected in the wild, and support the efforts of these companies and individuals in cultivating these beautiful plants.

the tropics, of living free of soil on the surfaces of other plants, especially trees and shrubs. This epiphytic habit enabled them to colonize the complex three-dimensional habitat of the tropical forest. The manifold opportunities available there for orchids to establish themselves eventually resulted in countless orchid species, each specialized for life in one of those niches.

Why are orchids such interesting flowers? It isn't simply their beautiful colors, because even the tiny white flowers are interesting. Perhaps it is their complexity. Some orchid flowers actually resemble animals. If seen from an appropriate angle, for example, the flowers of pink moccasin flower look like hummingbirds or hawk moths pollinating a different flower. Cranefly orchid (*Tipularia discolor*) is named after its resemblance to a cranefly, and the flower of yellow fringed orchid (*Platanthera ciliaris*) looks like a little face fringed (*ciliaris* means fringed) by a wild beard. Perhaps, like orchids' insect pollinators, we are tricked into seeing something other than a flower, and that is what we find so irresistible about them.

The Southern Appalachians are rich in terrestrial orchids, with common species blooming in progression from spring until fall. In addition to moccasin flowers, showy orchis and puttyroot bloom in spring. Small green woodland orchids and green adder's mouths come next in early summer, then cranefly orchids and rattlesnake plantains appear as summer wanes. In early fall, yellow fringed orchids are followed by ladies' tresses, which may bloom until frost. In addition to these common species, we have many uncommon ones as well. Most of their pollinators are known, but some are not.

For all their charm, orchids have yet another dark side: they parasitize fungi. The roots of our terrestrial orchids are intimately associated with the filaments of a soil fungus, which is enslaved by the orchid. The fungus delivers nutrients to the orchid from the breakdown of organic material in the soil, but the orchid provides little for the fungus. Even the dustlike orchid seeds, which have little stored nutrition, require such a fungus to germinate and grow.

## Morels: A Springtime Delicacy

April in Appalachia often abounds with them, but May is really the month for morels. Warm rains bring out this most prized of wild edibles. Morel mushrooms (that's mo*rel* not *mor*al) are among the easiest mushrooms to identify, but, as with anything you plan to eat, you should be absolutely certain of the identification before partaking.

Unlike the mushrooms that most people are familiar with, morels do not have gills. Instead, they produce spores in indentations on the outer surface of their pitted, or honeycombed, cap. The cap sits on top of a stalk and the whole mushroom can be up to six inches tall (Figure 2-5). The function of the mushroom (the fruiting body) is to disperse the spores of the fungus. Most of the fungal body, the mycelium, is underground, and the mushrooms we see above ground, like the flowers of plants, hold the reproductive cells where animals, wind, or rain can disperse them.

Morels are most often found growing in old apple orchards, where their mycelium grows on the dead roots of the apple trees, but they also occur under dying elms and in forests dominated by tulip trees or beeches and maples. They are easy to collect from burned areas, where the yellow mushrooms stand out conspicuously against the black earth. The mushrooms are produced from mid-April to mid-May in the Southern Appalachians, and a good, soaking rainfall is necessary to produce the heaviest crop.

The caps of yellow morels (*Morchella esculenta*) look like three-dimensional cones of golden or brown honeycomb, and are attached directly to a stalk that pushes them from the earth. Both the cap and the stalk are hollow and, when sliced, make beautiful rings for cooking. They occur through-

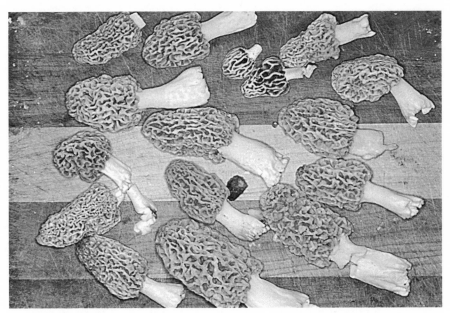

FIGURE 2-5
Morel mushrooms (*Morchella* sp.), prized edibles, have wrinkled caps and hollow stalks.

out North America but are especially common in the Appalachians and the Midwest. In order for the fungus to prosper, a cold period is required during the winter. Its scientific name refers to its esculent, or edible, properties — they have mild and earthy flavor with the slightest hint of apples. An annual festival in Michigan celebrates this tasty edible with a ninety-minute mushroom-gathering contest. One year's winner collected over 900 morels, which translates into ten mushrooms per minute! Minnesota has even declared the morel as its state mushroom.

Morel gatherers should be aware that in addition to the true morels, there are false morels (*Gyromitra* spp.), which are poisonous. False morels do not have a regularly shaped, honeycombed cap. Instead, the cap is convoluted and asymmetric. Any guide to mushrooms should have images of both false and true morels for visual comparison. Distinguishing between the two is easy, but be careful!

## Squawroot and Other Parasitic Plants

In spring, squawroot (*Conopholis americana*), one of our most unusual plants, thrusts upward through the leafy mat of the forest floor, pushing aside last year's brown leaves with its own emerging stem. Instead of a green stem and leaves, however, this plant is light brown, pale as a specter, for it lacks that one character that we associate with plants—chlorophyll.

Chlorophyll provides plants with a stunning capability. By capturing the energy of the sun, it allows plants to convert carbon dioxide into carbohydrates through the process of photosynthesis. If you lie out in the sun on a warm spring day, you might get a tan, but when you got hungry, you would have to go back to the kitchen for a sandwich. When a plant lies out in the sun, however, it is making carbohydrate food at the same time. All animals, including humans, must steal food molecules from plants. Only plants, along with a few single-celled organisms, can make their own food.

Because it lacks chlorophyll, squawroot (Figure 2-6) is unable to produce its own food. Like us, it requires a green plant to synthesize its food. Lacking a mouth, it uses its roots to form a permanent attachment to the roots of its host plant, parasitizing the host and absorbing food directly through this gall of tissue. Squawroot occurs throughout the East, where its oak tree (*Quercus* spp.) host can be found.

Squawroot looks like a cluster of corncobs sticking up out of the ground. Look carefully, and you will see that each "kernel" is a pale flower. The "cob" is the thick stem, which is covered by scalelike, triangular, brown flaps that are reduced leaves. Since the purpose of leaves is to provide a surface to intercept light for photosynthesis, plants without chlorophyll do not need leaves.

Squawroot is one of the favored foods of black bears, who emerge from their winter dens about the same time that the plant appears. Insects, espe-

FIGURE 2-6

The flower stalks of squawroot (*Conopholis americana*) emerge like corncobs from the ground in early spring (left). The plant parasitizes the roots of trees, invading the tissues to form a swollen knob or haustorium. The squawroot pictured on the right was being parasitized by aphids, the small white specks on the haustorium.

cially bumblebees, are attracted to the flowers, and they, at least, perform a service to the plant by pollinating it as they take away the food for themselves.

A related plant called beechdrops (*Epifagus virginiana*) parasitizes the roots of beech trees (*Fagus grandifolia*). Beechdrops have thinner stems than squawroot, with fewer and more widely separated flowers, but they are the same pale brown color. Both are members of the broom-rape family of plants. Beechdrops bloom in fall. Indian-pipe and pine-sap, which also lack chlorophyll, bloom in summer (see chapter 3). They are not true parasites of their tree hosts, relying instead on the exploitation of fungi to provide the needed nutrients.

## Honeybee Swarms: Colony Reproduction

Honeybees (*Apis mellifera*) are familiar insects to most of us. They are about 1/2 inch long, furry, with brown and yellow stripes on their abdomen (Plate

30). They forage for nectar, which they turn into honey, and pollen, which they feed their young. There are several related bees that produce honey, and they do so by concentrating the sugar found in nectar. It takes the nectar of about 2 million flowers to produce a pound of honey, and the 60,000 or more honeybees in a summer hive can produce an average of fifty pounds of honey every year. In a good year, my hives produce up to 300 pounds.

Honeybees, bumblebees, and many different types of solitary bees collect pollen in special baskets on their legs and use this pollen as a protein source for the developing young. In collecting this pollen, they often inadvertently pollinate the flowers they visit. Honeybees are important pollinators of most of our domesticated crops, and effective pollination of flowers is required in order to set the fruit we crave. Some of these crops, such as squashes, tomatoes, and beans, are native to the Americas (mostly Central and South America). Honeybees, however, originated in Europe. Honeybees are less important pollinators than our native bees for most of our native plants that have not been modified as crop-producing plants. The tongues of honeybees are generally rather short. In the open flowers of squashes and tomatoes, the short tongues are no handicap, and the bees easily gather abundant nectar while inadvertently pollinating the flowers. In many of our native wildflowers, however, the nectar is so deep that the short-tongued honeybees cannot reach it. These plants are adapted for our native long-tongued bumblebees, butterflies, and ruby-throated hummingbirds. In fact, honeybees often steal the nectar by cutting into the base of such long-tubed flowers as cardinal flower, so they bypass the fertile parts of the flower and thus don't pollinate it.

All the honeybees present in one hive are related as sisters. A single queen produced them all, for it is only the queen that mates and lays eggs. It is perhaps easiest to think about the hive as the greater organism rather than focus on the individual bees, for it is the hive that reproduces. When individual honeybees sting, which they do to protect the queen and the hive, their stinger is ripped from their body and they die. An individual's death does not matter as long as the hive itself is protected, for all the members of the hive share a complement of genes. Since the individual worker does not reproduce, the only way to reproduce its genes is to protect the queen. Each spring, the queen reproduces the colony.

As a beehive becomes crowded and the flow of nectar increases in the spring, the workers in the hive construct special large brood cells. The larvae in these cells are fed solely on royal jelly (secreted by the worker bees), and they grow into bigger individuals—female queens and male drones. These individuals leave the hive and mate. As this occurs, the original queen and most of her attendants leave the hive to find a new location to build. The newly mated queen then returns to her home hive, kills the remaining queens that are still in their cells, and starts laying a new set of workers. Some of the original workers (the new queen's sisters but not her offspring) remain behind to care for her and become part of the new colony.

The old queen and her attendants first cluster into a swarm (Plate 10) on a tree branch near the original hive. The queen is protected in the center of the swarm. Scouts from the swarm go out to locate a protected place to build a new hive. Beekeepers watch for these swarms and "hive" them by placing a hive directly under the swarm and then knocking the bees down into it. The bees recognize the smell of beeswax and nearly always remain in the new hive. This new hive can then be moved into a beekeeper's bee yard. As far as is known, once a new hive has been established by swarming, the workers do not get mixed up and return to the wrong hive.

Several pests threaten honeybees, and their presence causes expensive problems in the honeybee industry. Mites, hive beetles, and a host of fungal and bacterial diseases all live on the bees and can kill entire colonies when they invade. The newest threat is called colony collapse disorder, when whole hives die for no apparent reason. Pesticide residue on agricultural crops and in other areas of the environment is suspected as the major cause.

## Insects and Migratory Birds

So much happens in the spring months that the important relationships between organisms of the Appalachian cove forests are a little easier to see than they are in other seasons. For example, the wood warblers and other migratory birds arrive just as caterpillars are hatching (but see chapter 1). Warbler migration almost always peaks just before the trees leaf out, which is convenient for most birders; following these birds' rapid movement with binoculars becomes progressively more difficult as the leaves expand. Hungry caterpillars nibble on the fresh greenery but are themselves hors-d'oeuvres for the birds.

The Southern Appalachians see so many warblers moving through in spring and fall that it is impossible to describe them all. It is truly a delightful challenge to learn to identify them by sight and by song, and I, for one, must dust off my bird books each spring to refresh my memory. Two warblers define the southern mountains and are present throughout the summer months. If you visit the Southern Appalachians anytime from spring to fall, you have a good chance of seeing (or hearing) these birds. In fall, they migrate to Central and South America to overwinter with the other warblers.

Perhaps the warbler most often seen in the Southern Appalachians is the hooded warbler (*Wilsonia citrina*), with its loud call, bright colors, and bustling activity in the low rambles of rosebay rhododendron right at eye-level. The male is spectacular, surprising me with its beauty every time I see it. His bright yellow face is surrounded by a black hood, which covers the top of the head and slides down around the throat. His green back and yellow underparts complement the striking pattern of black and yellow on his head (Plate 11). As he flits about in the understory, he shows off for females by flashing his tail feathers. He pauses just long enough to make a loud call, often described as "wee-see, wee-see, wee-SEE-you." And we usually do!

Perhaps the warbler most often heard in the Southern Appalachians is the ovenbird (*Seiurus aurocapillus*), a secretive but loud-voiced summer resident. His emphatic "teacher, teacher, teacher" starts gently but crescendos to ring through the forest. This warbler, which walks on the ground, appears and behaves more like a tiny thrush than a fluttering, colorful warbler. Its back is brown, its breast is white with brown spots, and its head has a burnt-orange stripe along the crown. It is named for its nest, a little covered "oven," built on the ground. I can attest to its effective camouflage, for I have never seen a nest.

Like the ovenbird, the wood thrush (*Hylocichla mustelina*) is rarely seen but its voice echoes through the Appalachians in spring and summer. Related to robins and bluebirds, the wood thrush has a brown back and a white breast with brown spots and walks along the forest floor in the search for food. It migrates, as do the warblers, to follow its insect prey. Its call is a haunting, flutelike melody, to me the most melodious of all birds' calls. I will always associate it with cool foggy mornings in the mountains, accompanied by the dancing sound of creek water. At higher elevations, its relative, the veery (*Catharus fuscescens*), sings an equally ethereal song.

## Warbler Diversity and Decline

Over the last fifty years, warblers that nest in eastern North America but overwinter in tropical Central and South America have decreased in number, in some instances by as much as 70 percent. Several factors are probably combining to cause the decline. Habitat destruction leads the list, for it affects both nesting and wintering locations. When large areas of forest are destroyed, forest-nesting species obviously are eliminated. But even small-scale change from forest to open land has insidious effects. For example, when a large forest is cut into many smaller tracts by road building and development, the amount of "edge" increases relative to the amount of unbroken interior forest. Predators that are more common in open areas thus have more opportunity (more edges) from which to invade the remaining forest site.

Avian nest predators, including brown-headed cowbirds, blue jays, American crows, and common grackles, are more common in open sites than in mature forests. Cowbirds are nest parasites, whereas the other species eat eggs and nestlings. Nonavian predators, such as feral cats, rats, raccoons, eastern gray squirrels, and feral dogs, are also more common in open areas, and their populations have increased with the increasing human population. They consume eggs, nestlings, and adult birds. Feral cats, rats, and, to a lesser extent, feral dogs are especially problematic because they are nonnative invasive species that multiply at the expense of native wildlife, such as songbirds. Even well-fed pet cats regularly kill wildlife. If you have a pet cat or dog, keep it inside, where it is safer and so is the native wildlife!

## Caterpillars and Cuckoos

In May, the silky nests of eastern tent caterpillars (*Malacosoma americanum*) seem to be on every black cherry tree (*Prunus serotina*) between the mountains and the coast. The nests are bundles of silken threads almost always positioned in the fork of two strong branches. When they first become visible, they are only an inch or so in diameter, but as the caterpillars grow, so do the nests. To expand the nest, the caterpillars, which produce the silk from salivary glands in their heads and dribble it from their mouths, add new outer layers.

The tent nests not only protect caterpillars from bats, birds, and predatory insects, but it has also been suggested that they act as greenhouses. On cool spring nights, the temperatures can drop low enough that insects have difficulty moving and growing. A nest retains heat, keeping the caterpillars' metabolism high so that they efficiently digest their food and grow.

Caterpillars feed three times a day: in the early morning, at midday, and just after dark. Each time a worm leaves the nest, it lays down a silk trail so it can find its way back to the nest. Other caterpillars are attracted to the silken trails, and the best food sources usually have the thickest silk roads leading up to them because so many caterpillars have traveled them.

While the nests are the most visible aspect of the life cycle of tent caterpil-

One of the most famous insect naturalists, J. Henri Fabre, conducted a fascinating experiment with the pine processionary caterpillar, a European cousin of our tent caterpillars. Individuals of both species are hardwired to follow pheromones and silken trails. Under normal circumstances, the trails guide the caterpillars from the nest to food and then back to the nest, like Ariadne's thread in the labyrinth of the Minotaur. In Fabre's experiment, a foraging procession of caterpillars ascended the side of a large earthenware flower pot. Once they reached the rim and marched once around its perimeter, Fabre removed all traces of silk from the side of the pot they ascended. The caterpillars walked in an endless circular path for seven days, stopping only when the nights were cold enough to produce frost. He calculated that they walked at least 335 times around the rim for more than a quarter of a mile! Such experiments make clear that the success of insects is due, in part, to strength of instinct rather than intelligence.

lars, the eggs, caterpillars, pupae, and adults may also be encountered. The adult moths emerge in midsummer, mate, and lay eggs. They do not feed. The short-lived moths are stout and furry, about an inch long, and hold their wings up over their back. They are light brown, with two oblique white lines across the forewing. Although not present for long, they are abundant during their brief mating season.

After the leaves fall off the trees in autumn, the egg cases, which contain the overwintering eggs, become visible. The dark, shiny cases are about an inch long and completely surround twigs of the host tree, which is most often a black cherry. Each case can contain 200–300 eggs and there may be more than one case on each tree. Such unlucky trees may be completely defoliated by the caterpillars in the spring, when the eggs hatch.

Once the eggs from each egg case hatch, the sibling larvae congregate to construct their communal nest. They undergo five molts as they grow, and the old exoskeletons are left inside the nest along with their feces. This messy material is known as frass. If you have a nest nearby, you can keep track of each phase of the life cycle and watch as the larvae increase in size during the six weeks of their larval life. When they reach two or three inches in length, they are full grown and ready to pupate.

After the fifth molt, the caterpillars leave the nests and find individual places to spin a cocoon and pupate. The caterpillars are black and furry, with a white stripe down the middle of the back and blue dots along the sides.

They are visible crossing roads and sidewalks as they search for a protected place to spin their cocoons, such as under a window ledge. Pupation lasts for three weeks. The cocoon is tough and yellowish, and if you brush against it or shake it, yellowish powder is wafted into the air.

Fall webworms (see chapter 3) also build silken nests, but during summer and fall instead of spring. Unlike tent caterpillars' nests, webworms' nests are placed at the tips of branches and enclose the leaves on which the caterpillars are feeding. Fall webworms are rarely encountered until they come out of the nest to pupate.

Although most of our songbirds rely on insects during the spring as a major food source, few of them attack hairy caterpillars, and even fewer bother with the nests of eastern tent caterpillars. Nevertheless, one of our common, often heard but rarely glimpsed birds eats them with relish: the yellow-billed cuckoo (*Coccyzus americanus*). The cuckoo does, indeed, have a yellow bill, but that is rarely seen; instead, the long black-and-white banded tail identifies the bird. In my childhood, I learned to identify its croaking call as that of the rain crow, believing that it was calling for rain on hot summer afternoons.

## Blue Birds and Bluebirds

Every spring in early May, right around Mother's Day, I am treated to the sight of a striking solid-blue bird at my feeder. Surprising as it may seem, these small truly blue birds are not eastern bluebirds (*Sialia sialis*). Instead, they are indigo buntings (*Passerina cyanea*).

Indigo buntings are primarily seed-eaters and the males are completely blue, even on the belly. The females are brown, and only they incubate the eggs and feed the young. The brilliantly colored male does not visit the nest. Instead, he perches conspicuously on a nearby branch, singing loudly, and defends the nest from intruders. Indigo buntings are migratory, present in the Appalachians from late spring until fall.

Because eastern bluebirds do not eat seeds, they do not regularly visit bird feeders, although they will sometimes perch in nearby trees as other birds busily feed, perhaps attracted by all the bird activity. For most of the year, they eat only insects, but in winter, when insects are harder to come by, they include berries from plants such as dogwoods and hollies in their diets. Eastern bluebirds are not completely blue. The belly is white and the breast is rusty red. Females and young birds are slightly duller than the males but similarly patterned in blue, white, and red.

One other migratory blue bird that may appear at a feeder is the blue grosbeak (*Passerina caerulea*), but it is less common than the indigo bunting. The grosbeak has a thick, heavy bill like that of a cardinal and two wide bars of brown on its wings.

All three of these blue-colored birds are at home in meadows, orchards, and brushy edges of fields. Buntings and grosbeaks build open nests in bushes and shrubs. Bluebirds, however, are cavity nesters. They nest in trees with hollow centers but will readily accept birdhouses that are the right size.

A decrease in nesting sites and competition for existing sites with introduced species such as European starlings has reduced the bluebird population. One way to help bluebirds is to place nest-boxes for them along the edges of open fields. Their numbers appear to be increasing in most areas as they take to the man-made boxes. It is important that the box have the correct size of entrance hole (1 1/2 inches) and be at the correct height above the ground (5–10 feet). It is also recommended that no perching peg be placed on the outside, because while the bluebirds can easily cling to the box with their feet without one, starlings cannot.

You can also help bluebirds by eliminating pesticides from your garden. The birds will act as a natural insect control, especially of caterpillars. So put out seeds for buntings and grosbeaks and boxes for bluebirds, and then enjoy watching these beautiful birds all summer long.

## Migrations: The Return Route of Hummingbirds

April 15th may be tax day, but it is also the average date that ruby-throated hummingbirds (*Archilochus colubris*) return to western North Carolina each year. So, in advance of their return, I gladly take a break from tax returns to fill my hummingbird feeder.

Male birds arrive before the females and vigorously defend sources of food, including hummingbird feeders. In spring, natural sources of nectar are few and far between; native sources of nectar, such as jewelweed, cardinal flower, and bee-balm, are more readily available later in summer. If these plants are near your home, you will be treated to a show from not only the flowers but the birds as well.

On the East Coast, only the ruby-throated hummingbird is common. These birds winter in Central America and make the long migration each spring to their summer homes. The males have the ruby throat for which they are named, but the females and juveniles have white throats and bellies. Sometimes the ruby throat of the male appears black.

## HUMMINGBIRDS AND FEEDERS

Hummingbird populations have undoubtedly increased as the result of hummingbird feeders, but each feeder you buy seems to come with different instructions. It is not necessary to add red food coloring to the food, as some advise. The feeders themselves have enough red to attract the birds. It is also unnecessary to buy hummingbird food that has added protein and minerals, because the birds normally supplement their diet of nectar with protein-rich insects. On the other hand, additional protein and minerals probably don't hurt the birds either.

Most published recipes for hummingbird nectar call for one part sugar to four parts water to mimic the sugar concentration in the real thing, but I use a ratio of one cup sugar to two cups water. I boil the water, add the sugar, and pour the hot solution directly into the feeder to sterilize it; I let the solution cool to room temperature before hanging it back outside. I prefer the higher proportion of sugar because it is the sugar that provides the nutrition to the birds, not the water. The more concentrated syrup also means that any bird that sneaks past the watchful male to grab a quick mouthful will get more of the energy-rich sugar. Since April nights in the Appalachians are often cold, I would rather the birds were full of sugar that they can use to stay warm. The birds that visit my feeder seem to like the arrangement.

---

The females arrive from the wintering grounds shortly after the males. With their arrival, the males kick into high gear and begin to court the females. The courtship dance is a spectacular affair. The male flies in a huge semicircular arc, like a clock pendulum, with the female at the bottom of the swing. As the male passes by the female, he faces her with the feathers of his ruby throat fluffed up, and makes a peculiar whirring buzz. He may swing on this arc a dozen times, until the female moves away from her perch. She is always at the base of that pendulum arc, calmly evaluating his fitness as a mate on the basis of his aerobatic skill. Sometimes, as if in a desperate final attempt to convince her of his suitability, he buzzes back and forth like an angry bee right in front of her.

The birds often return to the same locations year after year. One year, when I was late with the feeder, a male bird hummed conspicuously in the exact location where the feeder had hung the previous summer. He even used the same perch as before and chattered loudly when I emerged with the feeder. I am convinced this was the same bird who left my feeder in September, flew all the way to Central America, perhaps where someone hung out another feeder, then returned to me a few months later.

The birds migrate, of course, because their predominant food source is the nectar of flowers, available only during late spring, summer, and early fall in the Southern Appalachians. In addition, their tiny bodies lose heat rapidly, necessitating a warm environment, so they follow the heat as well. While they focus on flowers for food, like most other birds they also eat insects, especially during their breeding period. Insects and other animals are composed of about 70 percent protein, whereas plants typically have less than 5 percent. Protein is needed for fast-growing nestlings to grow feathers, reach adult size, and begin to fly, all in a few short weeks. As small birds, they eat small insects, and most frequently dine on gnats and midges that they pluck from the air. Next time you notice a hummingbird hovering in midair, look closely to determine whether it is actually nabbing gnats.

## Spring Peepers

Just as birds rely on the warming temperatures of spring to encourage insect populations, frogs and salamanders also depend on insects for their food. Of all the frogs that vocalize during spring in the Southern Appalachians, spring peepers (*Pseudacris crucifer*) are probably the best known. To many residents, they are the harbinger of spring.

These little tree frogs are about an inch long, with the swollen toes typical of most tree frogs. They are tan, with two dark lines that form an X across their backs (Plate 12); "crucifer" refers to this cross. Peepers are common not just in the mountains, but throughout the Southeast. Along the coastal plain, they often breed in winter. They congregate near shallow ponds to lay their eggs, but, like other tree frogs, disperse into underbrush and low trees when they are not breeding. I frequently come across them in my garden in the late spring and summer.

The peeps of spring peepers are rapid calls, occurring about once each second, and so clear and high that they sound like bird calls. Rarely do they call alone, however. Under most conditions, but especially during the moist spring nights that are most conducive to reproduction, a chorus of peepers is heard. In fact, when they are most active, the peeps run together into a loud, continuous, high-pitched scream that can be deafening. One evening when I walked quietly up to a breeding pond (with my hands held over my ears), they did not hush their amorous calls until I was standing nearly upon them!

## Climatic Conditions in Spring

In terms of weather, March is the most variable month of year in the Southern Appalachians. There can be snow, freezing temperatures, high winds, or warm spells. The average high temperature in Great Smoky Mountains National Park in Gatlinburg, Tennessee, is 61°F, and the average low is 33°F. These averages, however, are calculated over a thirty-year period. Any day's temperature can differ tremendously from those averages. On average, 5.6 inches of precipitation fall over ten days, and 1.2 inches of snow fall during the month. The average day length for the month is 12 hours in Gatlinburg. The spring equinox is March 20 or 21.

April is a calmer month, adhering more to springlike average temperatures. Its average high is 69°F, its average low is 39°F, and an average of 4.4 inches of rain falls over eight days. Frosts are still common, and an average of 0.9 inches of snow falls during the month. The average day length is 13 hours.

May's average high temperature is 76°F, and its average low is 49°F, with an average of 5.6 inches of rain over ten days. The average last frost date in Gatlinburg is May 1, meaning that 50 percent of the last frost dates fall before and 50 percent fall after this date. The average day length is 14 hours.

Many of the animals and plants highlighted in this chapter remain visible through the summer. Animals that depend on environmental temperature to warm their bodies, especially insects, amphibians, and reptiles, become even more active. Small spring flowers are overgrown by the big plants of summer as they transform into seeds, and the spring ephemerals fade away altogether, replaced by the rangy flowers of summer.

# Cycles of Summer
## June, July, August

3

June, July, and August in Appalachia bear witness to wildflowers of incomparable beauty. The summer sun fires the blues, reds, and yellows to an intensity that almost singes the eye. Color seems to attract color as ruby-throated hummingbirds, golden honeybees, and countless colorful butterflies are drawn to the flowers as surely as lips to a kiss. Frequent heavy rains, moist, humid air, long hours of sunlight, and warm temperatures produce a profusion of lush plant growth.

## Flowers in Myriad Forms: Jewelweed and Dodder

The stunning flowers of jewelweed (*Impatiens capensis*) hang like orange-spotted gems among the silky green leaves. Each inch-long flower resembles a pouch, or cornucopia; its open mouth narrows into a long, slender nectary that coils under the blossom (Plate 13). Only pollinators with long and flexible tongues can reach the nectar lodged deep in the flower's interior, and ruby-throated hummingbirds are one of its most frequent visitors. Sphinx moths, bumblebees, and some large butterflies also visit the flowers. In the Southern Appalachians, orange or spotted jewelweed is most common, but at higher elevation and father north, *I. pallida* is more frequently encountered. Compared to its southern relatives, it is truly pallid, its flowers a light yellow in color.

Although jewelweed's flowers certainly contribute to the plant's name, it may also get its name from the seeds of the plant. While the seed is initially green or brown, depending on how ripe it is, once its coat is peeled away, a Carolina blue gem appears, as a tiny sapphire might slide out of a velvet bag.

Jewelweed is an old-timey remedy for poison ivy exposure (see chapter 2). The plants are succulent, with thick, juicy stems. If the plants' sap is rubbed immediately on skin that has been exposed to poison ivy, it is reputed to be an antidote. (I cannot personally confirm or deny its efficacy as an antidote.) The juicy sap is as soothing as aloe if applied to an itchy poison ivy rash, or even bug bites and sunburn. Clinical studies so far support jewelweed juice only as an anti-itch remedy, but folk medicine enthusiasts claim the plant provides many other benefits.

Perhaps it is this seed, not just the flower, that is the jewel. Yet a third possible basis for the name jewelweed is that water beads up on the hydrophobic leaves so that raindrops dangling from the leaf sparkle like diamonds; alternatively, if the leaf is held underwater, the air bubbles trapped on its underside look like tiny jewels.

The juxtaposition of the terms "jewel" and "weed" is certainly odd, but the plants really are weedy. They grow up to four feet high, are lanky, and take up a lot of space wherever they grow. And they grow anywhere there is moisture, in damp areas around ponds, for example, or in ditches along shady roadsides, where their orange jewels of flowers wink among the rambling plants.

The plant is also known as touch-me-not, because of the way the seeds are released from the plant. Once the seed pods are ripe, they explode with the lightest touch of finger or feather, and the seeds may be thrown several feet from the plant. These seeds are easy to collect and plant, and once the plants are established, they appear everywhere. I generally leave them to fend for themselves around the mulched bases of trees in my yard, where the trout lilies emerge in spring and are gone by the time jewelweed has grown.

Jewelweed also has a fantastic ability to set seeds. Annual plants such as jewelweed must set seeds every year to ensure their continuation, but jewelweed is unusual in that its seeds do not persist as a seed bank in the ground. Instead, they all germinate and grow in the spring following their production. It is imperative that seeds are produced each year, no matter what the growing conditions and whether or not a pollinator visits. The large, attractive flowers of jewelweed attract pollinators, but jewelweed produces small flowers as well that do not require a pollinator, that is, they are self-fertile. In terms of energy, these small flowers are less expensive to produce. When jewel-

FIGURE 3-1

When the parasitic plant dodder (*Cuscuta* sp.) attaches itself to its host plant, pegs of tissue invade the host's stem.

weed grows in marginal habitats, such as drier sites, it typically produces more of these self-fertile flowers. Its limited energy reserves can thus be redirected from expensive flowers that may or may not be cross-pollinated to cheaper, self-fertile flowers that will definitely produce seed. To ensure a successful pollination of the cross-pollinated flowers, the flowers dangle from a long pedicel; because the wobbling flowers require a longer visit from the pollinator to extract nectar, more pollen is deposited and the chance of cross-pollination is improved.

Frequently twined among jewelweed are the pretty vines of morning glories (*Ipomoea* spp.). Like jewelweed, most morning glories are annuals. Their vines struggle up and soon overtop supporting plants, and one morning glory in particular has taken this habit to the extreme. Dodder (*Cuscuta* spp.) is a parasitic plant of the morning glory family. It begins life like any other vine, germinating and growing quickly up a nearby plant, but as it grows, it sends out pegs of tissue (Figure 3-1) that invade the stem of the host plant and suck off its nutrients. Its own roots eventually die because it obtains all its nutrients from the host. By late summer, the yellow vines droop like masses of cooked spaghetti over the host plants (Plate 14). Look closely at that spaghetti, however, and you will see that it is covered with 1/4-inch-wide flowers. Because it is a parasite, drawing its nutrients from a host, it has dispensed with chlorophyll and large leaves, as has squawroot (see chapter 2), another parasitic plant. All you will see is the yellowish stem and tiny white flowers of the plant.

## Summer's Divine Robes: Cardinal Flower

In every season, one characteristic plant or animal seems to capture my heart as the quintessential expression of nature's cycles. I know summer is giving way to fall when cardinal flower (*Lobelia cardinalis*), surely one of our most spectacular wildflowers, blooms. Three-foot-tall, scarlet stalks accent pond edges and meadows, luring butterflies and hummingbirds to the visual and culinary feast (Plate 15). Each individual flower on the flower stalk is a long tube, about two inches in length, with white-tipped stamens arching upward and sticking out beyond the corolla. The lower petal is divided into three large lobes, and an upper petal divides into two thinner ones. As with the northern cardinal, the brilliant red bird of our yards and forest edges, cardinal flower is named after the scarlet robes worn by cardinals of the Roman Catholic Church. Several other lobelias also occur in the Appalachians, but they have blue, purple, or white flowers.

One afternoon, my husband and I watched a ruby-throated hummingbird pause at the blood-red flowers in our yard, each time bumping his forehead against the stamens and transferring pollen to the next flower. The stamens were sturdy enough to ruffle his feathers, leaving a tiny cowlick on his forehead. On another day, a pipevine swallowtail butterfly moved up one stalk of the ruby flowers and down another, tasting each sweet offering and transferring pollen as it moved.

Cardinal flower grows most profusely where it receives direct sunlight but still gets plenty of water. In the wintertime, the tall flower stalk dies back, leaving a basal rosette of green leaves on the surface of the ground. This leafy rosette must remain free of debris, dead leaves, and competing vegetation or the plant will die. Because it can tolerate wet and soggy conditions that most plants cannot, cardinal flower often grows along stream banks or below the water line in seasonally flooded ponds. The spring floods and high water wash away decaying leaves and inhibit competitors, leaving unscathed the cardinal flower seedlings that grew from seeds shed in the fall, when water levels are lowest.

As each pollinated flower withers, the fruit enlarges into a swollen bulb that encloses the seeds. When the bulb dries out, the hundreds of brown seeds inside are ripe and easily scattered by wind. They are so small that they float on the surface of the water, eventually lodging against a sandy bank, where they take root. Once established, new plants constantly appear.

## Ginseng and Yellowroot:
## Uncommon and Common Medicinal Plants

Ginseng (*Panax quinquefolius*) is a well-known medicinal plant that is becoming increasingly uncommon, mainly from overcollection. It grows in rich, fertile coves and must grow for several years before becoming reproductive and producing bright red berries (Plate 16). Each compound leaf is made up of five leaflets, and usually two or more sets of leaves are present in berry-bearing plants. The ten or so berries are held up on a stem about eight inches from the ground and are produced in late summer.

Humans are both a major source of seed dispersal and of plant removal. Because the root is the valuable part of the plant, the entire plant is killed by the harvest. Harvesters use the bright red berries to locate the plant in the forest just as do the native animals, such as wild turkeys, that disperse the fruits. The legal harvest season for the root is intended to coincide with the ripening of the fruit, and ethical harvesters plant the seeds as they harvest the parent plant. Replanting the seeds should ensure a sustainable harvest — as long as the seedlings grow and set seeds themselves. It is of the utmost importance, then, that the seeds are fully ripe before the plant is taken so that the seeds germinate; in some areas the legal harvest season occurs before seeds become viable. It is also important that the plants be allowed to grow to maturity before they are removed; a population cannot be sustained if juveniles are removed before they reproduce.

Ginseng grows in rich, fertile forests, and it once thrived across eastern North America. When the Asian species became rare, our American ginseng was collected for export to China for use in traditional medicine. It is famed for preventing fatigue, promoting alertness and mental ability, promoting longevity and vigor, and even preventing children from having bad dreams. But this fame has come at a price — ginseng is now rare, and many populations are still in decline even though collection and trade of ginseng are now regulated in most states.

Yellowroot (*Xanthorhiza simplicissima*) is a medicinal plant that was important to the Cherokee. It remains common most likely because its fame as a medicinal has not spread as widely as ginseng's. Like goldenseal (*Hydrastis canadensis*), yellowroot is used as a general tonic to improve health. It is also used to clean wounds, encourage healing, and treat digestive problems, among other things. It was also extremely important to Cherokee as a yellow dye. Traditionally, the Cherokee used yellowroot to produce a yellow dye, bloodroot to produce an orange-red dye, and black walnut to produce a brown dye used in their basketry.

Yellowroot is a small, woody shrub that can grow up to three feet tall. It grows along shady stream banks. Like ginseng, it has compound leaves, but the leaves are narrower and composed of more leaflets. They arise from the tip of the rarely-branched, grayish stem. In early spring, clusters of small purple flowers bloom before the leaves have fully expanded (Plate 17). By summer, these flowers have transformed into small greenish fruits. The underbark of the stems and the roots are, indeed, yellow.

## Insectivorous Plants: Sundews and Pitchers

At sites where rivulets trickle and splash along the rocky outcrops of the Blue Ridge Parkway, unique communities of plants flourish. If you take the time to get out of your car and look, you may be surprised at what you find. My favorite location is the Wolf Mountain Overlook near mile marker 425. Here, in a sort of vertical minibog, carnivorous sundew plants are interspersed with small, green woodland orchids, turtlehead, saxifrage, mosses, and rare flowers, such as the beautiful grass of Parnassus. These gems bloom in August, and their flowers linger into September.

The round-leaved sundew (*Drosera rotundifolia*) is common in this seepage cliff community. The entire plant is rarely more than two inches in diameter. The small, round, 1/4-inch diameter leaves are covered with hundreds of glandular tentacles, each resembling a reddish lollipop that exudes a sticky droplet of sweet liquid (Plate 18). This confection is irresistible to small insects, especially ants. A foraging ant investigating this attractive leaf soon becomes mired in the tentacles. The futile struggle of the ant only causes more tentacles to bend toward, and then around, the prey, pinning it in a final sweet embrace. Digestive enzymes are soon released and the hapless ant is reduced to a nutrient broth, which is absorbed through the surface of the leaf. Darwin was intrigued by sundews, and investigated their ability to distinguish between food items such as ants and meat and nonfood such as sand grains or moving twigs that mimicked the struggles of an ant. The plants can, indeed, distinguish the difference, probably by detecting chemicals in the food.

Insect-eating pitcherplants, named because they hold water like pitchers, also occur in the Appalachians. A cluster of these pitchers, which are modified leaves, grow in a rosette, and in summer a large maroon or yellow flower develops on a foot-tall stalk. The colorful flowers are designed to attract insect pollinators for the sake of sexual reproduction, which results in seeds. The large, conspicuous, and often colorful leaves also attract insects, but for the nefarious purpose of eating them. The pitcher leaves exude sweet nectar from glands on the lids, pitcher rim, and just inside the mouth of the pitcher. When a hungry insect enters the pitcher, it slips down the lining of the pitcher, which is carpeted with myriad stiff hairs that point downward toward a pool of liquid in the bottom of the vessel. The prey falls into the pool below, drowns, and decays. The decay is facilitated by digestive enzymes released from bacteria in the pool; pitcherplants do not themselves secrete digestive enzymes. Nutrients from the digested animal are absorbed and used by the plant. Both the rocky habitat of round-leaved sundews and the acidic sphagnum bogs of pitcherplants are very low in nutrients, so insectivorous plants acquire those nutrients, especially nitrogen, from the bodies of animals they digest.

Different species of pitcherplants have leaves that are either upright, like organ pipes, or splayed horizontally across the ground, like cornucopias, but all species function as pitfall traps, as described above. In the Appalachians, three species occur naturally and others have been introduced. Both the mountain sweet pitcherplant (*Sarracenia jonesii*) and the green pitcherplant (*S. oreophila*) are rare endemics of the Southern Appalachians. Their elegant upright pitchers stand about a foot high and are up to two inches wide at the top (Plate 19). The purple pitcherplant (*S. purpurea*) is more widely distributed both in the mountains and along the coastal plain, where most other species of pitcherplants occur. A subspecies, the Southern Appalachian purple pitcherplant (*S. purpurea var. montana*), has been identified as another specialist of our region. Both purple pitchers have cornucopia-like pitchers that are short and fat, about six inches tall and three inches wide (Plate 20). Unlike the other mountain species, these pitchers lack the overhanging lid. Because the pitchers are wide open and rainwater dilutes the digestive enzymes, they each create a protected mini-pond suitable for colonization by specialized aquatic animals. One of these is a group of nonbiting mosquitoes that breed only in pitcherplants. The mosquitoes lay their eggs in the pitchers, and the larvae develop there before transforming into the flying adults. One species, *Wyeomyia haynei*, breeds in our purple pitchers, while *W. smithii* occurs in the more northern populations of *S. purpurea var. gibbosa*.

Why are so many species of pitcherplants rare? Both the mountain sweet pitcher-plant and the green pitcherplant are federally endangered, as are some species that occur outside the Appalachian region. They are rare primarily because their natural bog habitat, and mountain bogs in particular, are rare or are becoming rare. Because boggy areas are generally viewed as nonproductive, many have been drained or filled in and converted into agricultural fields or housing projects.

Fire suppression can also eliminate bogs. Because of the shape of their leaves, pitch-erplants are not very effective at trapping sunlight for photosynthesis. To offset this handicap, they prefer, and thrive in, sites that receive high levels of sunlight. In the absence of fire, boggy areas and pond edges eventually will be colonized by shrubs and trees, which sooner or later will shade out the pitchers. Thus in most areas, the persistence of the pitcherplant community depends on periodic natural fires to de-stroy the competing shade plants.

A final challenge to the survival of wild communities of pitcherplants is overcol-lecting by enthusiasts and profiteers. Pitcherplants are botanical treasures, and the ensuing demand, especially for rare species, encourages poaching. Fortunately, tech-niques for raising pitchers from seed or tissue culture have improved, and several botanical groups are involved in projects to restore depleted pitcherplant habitats.

---

Moths are a common food for the purple pitchers; their bodies sink, but their detached wings remain afloat. Ants are by far the most common food for the mountain sweet pitchers, although yellow jackets and other wasps are also frequent prey. Carcasses in the bottom half of a pitcher are often packed as tightly as tobacco in a cigarette paper.

## The Heaths: Mountain Laurel, Rhododendron, Sourwood, and Pinesap

Visitors flock to the Southern Appalachians in June and July not only to escape the heat of the low country but also to enjoy beautiful flowering plants. Many plants that define these mountains belong to the heath fam-ily (Ericaceae): mountain laurel (*Kalmia latifolia*), great or rosebay rhododen-dron (*Rhododendron maximum*), and Catawba rhododendron (*R. catawbiense*) are probably the three most widely appreciated of these plants for their magnificent flower displays. All have evergreen leaves and can reach about twenty-five feet in height, but the leaves of mountain laurel are about two inches long and those of the other two are six or more inches long. The leaves of rosebay rhododendron and Catawba rhododendron are difficult to distinguish from each other, but their flowers are pink (or white) and purple, respectively. Expanses of Catawba rhododendron put on quite a display at Craggy Gardens, Roan Mountain, and many other sites. It is a Southern Appalachian endemic and usually occurs at higher elevations than either mountain laurel or rosebay rhododendron.

The flowers of mountain laurel come in shades of pink or white and are shaped like upside-down ballerina tutus, with the petals fused together. Each of the ten pollen-bearing stamens arches outward and anchors its tip inside the rim of the flower in a tiny pocket until the pollen has matured. Then, when an insect lands on the flower and touches the stamen filament, its tip pops loose from the flower, snaps over, and smacks the insect on the back, dusting it with pollen. Once tripped, the stamens remain curled (Plate 21). When the insect visits another flower and brushes against the female stigma, it transfers its load of pollen. You can imitate the insect by sticking your finger into the flower of a mountain laurel. Poke the spring-loaded filament and it should snap forward to leave a yellow dusting of pollen on your fingernail. If you play this game, however, finish it by transferring pollen to another flower's stigma. If the flower is on another plant, the likelihood of successful seed set is greater.

The largest and latest blooming heath is sourwood (*Oxydendrum arboreum*), which typically grows up to forty feet high and twelve inches in diameter, and occasionally even larger. Most often a component of the understory, it can be identified by its dark, blocky bark and by the leaning or arching growth form of the tree itself. Like other heaths, it rarely grows straight up but rather bends in odd shapes that add interest to the forest canvas. If grown as a specimen tree, and it doesn't have to compete for light with other plants, it grows straighter. Sprays of white flowers, the size and shape of blueberry flowers (another heath), are clustered at the branch tips. Some years the blossoms are so plentiful that the trees glow white in the otherwise green summer forest. In such years, an abundance of high-quality sourwood honey is produced.

In drier years, the leaves of sourwood are decimated by fall webworms (*Hyphantria cunea*), which build their silken nests in the tips of the branches. The caterpillars enclose the leaves of the tree in silk, securely feeding within their retreat. Sourwood is their favored tree, followed by black cherry and a few others. The nests are sometimes confused with those of tent caterpillars (see chapter 2), but tent caterpillars build their nests in the forks of branches during the spring months.

One reason that heaths are so common in the Southern Appalachians is that they thrive in soils with high acidity and few nutrients. Like carnivorous plants, they have a "trick up their sleeves" that allows them to exploit locations where other plants do not thrive. The roots of heaths have a unique relationship with mycorrhizal fungi. The high acidity (low pH) of soils in which these plants grow inhibits the normal bacterial breakdown of humus and release of nutrients. The fungi produce an enzyme, however, that breaks down soil humus and releases nutrients, especially nitrogen, to the plant. The fungi enable heaths to exploit and prosper in an otherwise marginal environment where they don't have to compete with other plants. In fact, once heaths such as great and Catawba rhododendrons gain a foothold, they often grow so densely as to shade out other plant competitors. Similarly, sourwood and trailing arbutus, another heath (see chapter 5), successfully grow on road cuts in poor subsoil but where their leaves have abundant sunlight and little competition.

Some heaths are so dependent on their fungi that they can't survive without them. The common pinesap (*Hypopitys monotropa*) and Indian pipes (*Monotropa uniflora*), as well as the rarer sweet pinesap (*Monotropsis odorata*), are heaths that can't produce their own carbohydrate food because they lack chlorophyll. They rely solely on their mycorrhizae to acquire inorganic nutrients, such as nitrogen, *and* carbohydrate sugars. The mycorrhizal network of filaments in the soil links the roots of these nonphotosynthetic heaths with the roots of unrelated photosynthetic trees. The fungi parasitize the tree roots, withdrawing sugars from the tree and delivering them to the heath.

Indian pipes are nodding white plants about six inches high with a single flower per stem that arise in clusters from the forest floor during the summer months. When the flower is pollinated, primarily by bumblebees, it turns upright. In winter the dried brown stalks are a common sight.

Pinesap (Plate 22) is similar, but each stem bears several flowers, and the stems are either yellow or reddish. Scientists suggest that the different colors indicate different species, for they usually bloom at different times (and therefore would be genetically isolated from each other). The yellow form blooms early in summer, but the red one blooms much later, even into fall. Both plants are common in acidic forests of oaks and pines or under hemlocks and rhododendrons, and they grow in dense shade because they do not require light.

PLATE 1. Serviceberry (*Amelanchier arborea*) trees bloom early in spring, often appearing as white patches in an otherwise gray forest.

PLATE 2. The flowers of Carolina silverbell (*Halesia tetraptera*) resemble white bells. They are filled with yellow stamens and have a central pink stigma.

PLATE 3. The flowers of oaks, such as those of the white oak (*Quercus alba*) pictured here, occur in clusters. The male flowers hang down in greenish yellow catkins at the juncture of new green growth and the previous year's brown twig. The tiny red female flowers occur at the base of the leaves with the stem.

PLATE 4. The flowers of tulip trees (*Liriodendron tulipifera*) are reminiscent of tulips. Multiple pistils are borne on a central cone, from which the flowers will eventually produce dozens of seeds. Multiple stamens surround the pistils and align with the orange stripe in the petals, attracting the insects that readily pollinate the flowers.

PLATE 5
The burgundy (and sometimes white) flowers of wake-robin (*Trillium erectum*) are held on stalks above the three leaves.

PLATE 6
The jack-in-the-pulpit (*Arisaema triphyllum*) can be confused with a trillium because its leaves also occur in threes, but the unusual "flower" is distinctive.

PLATE 7. Oconee bells (*Shortia galacifolia*) blooms in early spring. The leaves of Oconee bells look similar to those of galax but are shinier and more obviously veined. Each white flower is borne on a single stalk.

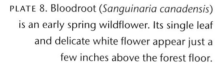
PLATE 8. Bloodroot (*Sanguinaria canadensis*) is an early spring wildflower. Its single leaf and delicate white flower appear just a few inches above the forest floor.

PLATE 9. Pink moccasin flower (*Cypripedium acaule*) is pollinated when a bumblebee becomes trapped inside the flower and escapes by passing under the pollinia.

PLATE 10. A swarm of honeybees (*Apis mellifera*) protects the queen at its center while scouts search for a new home. The swarm is the hive's method of reproduction.

PLATE 11. The male hooded warbler (*Wilsonia citrina*) is magnificently colored and often seen because he frequents rhododendron shrubs rather than treetops.

PLATE 12. Spring peepers (*Pseudacris crucifer*) are tan-colored tree frogs, about an inch long and marked with a darker X on the back.

PLATE 13. Jewelweed (*Impatiens capensis*) flowers are orange gems. Both the orange flowers and green seed pods (above the flower) are present in the same season.

PLATE 14. Dodder (*Cuscuta* spp.) is a parasitic plant that droops over its host.

PLATE 15. Cardinal flower (*Lobelia cardinalis*) is a magnet for humming-birds and butterflies, such as the pipevine swallowtail (*Battus philenor*) pictured here. The plants typically grow around ponds and streams.

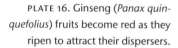

PLATE 16. Ginseng (*Panax quin-quefolius*) fruits become red as they ripen to attract their dispersers.

PLATE 17. Yellowroot (*Xanthorhiza simplicissima*) flowers in early spring as the leaves are unfurling. Along with ginseng, it was an important medicinal plant for the Cherokees.

PLATE 18. The round-leaved sundew (*Drosera rotundifolia*) occurs on moist rock faces and captures insects with the sticky pink extensions on its leaves.

PLATE 19. The rare mountain sweet pitcherplant (*Sarracenia jonesii*) occurs in bogs of the Southern Appalachians. The tall modified leaves produce a sweet fluid to attract and then drown insects.

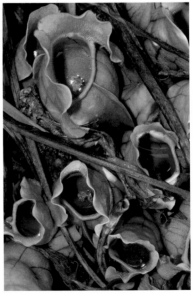

PLATE 20. The Southern Appalachian purple pitcherplant (*Sarracenia purpurea var. montana*) is a subspecies of a more widely distributed pitcherplant. The pitchers are prostrate on the ground and open widely above.

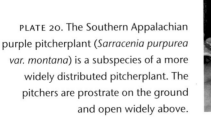

PLATE 21. The flowers of mountain laurel (*Kalmia latifolia*) are shaped to disperse pollen onto the insects that visit them. The stamens of the flower on the right are all still held in the pockets of the flower, but the stamens of the flower on the left have been tripped.

PLATE 22. Pinesap (*Hypopitys monotropa*) is pale pink or yellow because it lacks chlorophyll. Clusters emerge from the shady forest floor.

PLATE 23. The striking fruits of doll's eyes (*Actaea pachypoda*) are poisonous to humans.

PLATE 24. The pink fruits of hearts-a-bustin' (*Euonymus americanus*) open to reveal the attractive red seeds.

PLATE 25. Stinkhorns, such as the Ravenel's stinkhorn (*Phallus ravenelii*) pictured here, attract gnats and flies with their strong smell.

PLATE 26. Some fungi, such as this beetle cordyceps mushroom (*Ophiocordyceps melolonthae*), invade the tissues of a host organism, which, in this case, is a beetle pupa.

PLATE 27. Sulfur shelf
(*Laetiporus sulphureus*) is an
edible bracket fungus that
grows on dead trees.

PLATE 28. Bird's nest fungi, or splash cups
(*Cyathus striatus*), look like tiny nests with
eggs inside. These "eggs" are spore masses
that are knocked out of the cup by raindrops.

PLATE 29. During warm, rainy periods in spring, slime molds, such as the
common dog vomit slime mold (*Fuligo septica*) pictured here, rapidly appear,
release spores, and disappear again.

PLATES 30 and 31. Drone flies (*Eristalis tenax*) closely mimic honeybees (*Apis mellifera*), and the few differences between the two are hard to see in a living, moving fly. The four wings of the bee (left) are held over the back, while the two wings of the fly (right) are held at an angle from the body. The antennae of the bee are long, while those of the fly are short. The huge eyes of the fly take up most of the head and touch each other at the top, but the bee's eyes are on either side of its head. The mouthparts of the fly are tubular and used only for sucking, but those of the bee are larger and used for chewing. Bees also possess a tongue with which they can lap up nectar.

PLATE 32. The squash borer moth (*Melittia cucurbitae*) is a serious garden pest that mimics a wasp. Its hind wings are clear, and it flies very fast.

PLATE 33. When disturbed, the caterpillar of a tiger swallowtail butterfly (*Papilio glaucus*) rears up and extends a red, Y-shaped gland. With the accompanying eyespots, it looks like the head of a snake.

PLATE 34. This adult dragonfly has just emerged from its larval skin, to which it clings while waiting to unfurl its wings by pumping blood into them.

PLATE 35. The goldenrod crab spider (*Misumena vatia*) is so well camouflaged that it is difficult to see until it has captured its prey, like this unlucky bumblebee (*Bombus* sp.).

PLATE 36. The black and yellow millipede (*Boraria stricta*) produces a strong cyanide odor when disturbed.

PLATE 37. A white color morph of the common gray squirrel (*Sciurus carolinensis*) occurs in Brevard, North Carolina, and elsewhere. It is not an albino, for it has dark eyes and a gray stripe down the center of its back.

PLATE 38. The most visible salamander in the Southern Appalachians is the red eft (*Noto-phthalmus viridescens*). During the eft stage, the salamander is bright orange and terrestrial.

PLATE 39. The five-lined skink (*Plestiodon fasciatus*) is an active lizard of the forest. Juveniles have a blue tail that distracts predators.

PLATE 40. The copperhead (*Agkistrodon contortrix*) has a large head and is strikingly patterned in tan and brown.

PLATE 41. The nonpoisonous northern water snake (*Nerodia sipedon*) has a narrow head, reddish brown bands, and a gray body. It is often mistaken for a copperhead or water moccasin but is not poisonous.

PLATE 42. The leaves of red maple (*Acer rubrum*) demonstrate that as chlorophyll degrades, other pigments such as anthocyanin become visible. Chlorophyll is reabsorbed through the petiole at the leaf base.

PLATE 43. The purple flowers of stiff gentian (*Gentianella quinquefolia*) occur along the open areas of the Blue Ridge Parkway, especially near Devil's Courthouse Overlook.

PLATE 44. Nodding ladies' tresses (*Spiranthes cernua*) prefer open areas that are constantly moist.

PLATE 45. On male monarch butterflies (*Danaus plexippus*), thin black lines swell into a thickened spot on each hind wing.

PLATE 46. The width of the brown band on woolly bear caterpillars (*Pyrrharctia isabella*) is determined by age, not winter's severity.

PLATE 47. Nursery web spiders (*Dolomedes vittatus*) guard their egg case and spiderlings. They tend to occur around water, as does cardinal flower (*Lobelia cardinalis*), in which this spider has built her nest.

PLATE 48. During winter freezes, groundwater freezes and squeezes up through clay soils to provide this display.

PLATE 49. This Asian multicolored ladybug (*Harmonia axyridis*) is feeding on hemlock woolly adelgid (*Adelges tsugae*), both of which are introduced species in the Southern Appalachians. The ladybug is identified by the black letter M on its otherwise white "neck."

PLATE 50. Dead hemlock trees (*Tsuga canadensis*) stand out conspicuously among the summer greens of deciduous trees. This photograph was taken near Graveyard Fields along the Blue Ridge Parkway.

## Interdependence of Plants and Pollinators: Yucca

One of the most famous relationships between a plant and its pollinator is that of *Yucca* and its moth. The first study of this relationship was published in the 1890s, and it has been an ongoing area of fascination and scientific inquiry, with another pioneering study published in 1992. Both the plant and the moth are dependent on each other for their reproduction. Yucca moths pollinate yucca plants, ensuring development of seeds, and yucca plants are the sole nursery for the eggs and larvae of yucca moths. If either plant or insect disappear, both are doomed.

The leaves and flower stalk make *Yucca* easy to identify. The large, pointed leaves, which grow up to 2 1/2 feet long, are rigid and swordlike, persisting for many years on the plant. They radiate out from a central base, from which a single flower-stalk, four feet tall or more, elongates in spring. The 1 1/2-inch-wide flowers are creamy white, and the tall flower-stalk bears many of them, which open during the summer months. Their scent is strongest at night to attract their pollinator moths. In the Appalachians, *Yucca filamentosa*, identified by the stringy filaments that hang off the leaf edges, is the common species.

The yucca moth (*Tegeticula yuccasella*) is a pale, creamy white, similar in color to the flowers of *Yucca filamentosa* (Figure 3-2), and about 1/2 inch long. Female moths shelter in the flowers during the daytime but become active around dusk as they deposit their eggs and pollinate the flowers. First, the moth climbs one of the sturdy stamens and scrapes off pollen with her mouthparts. When she has a ball of pollen large enough to wedge comfortably between her mouth and first pair of legs, she flies off to a second flower on another plant. By choosing to fly to another plant, the moth ensures that the plant is cross-fertilized and outbred.

The moth evaluates each flower's ovary by using sensory organs scattered over her legs, ovipositor, antennae, and other body parts, ultimately settling on one flower that is in the right stage of development and does not contain other moth eggs. She then bores a hole into the ovary with her ovipositor and lays one egg. After laying the egg, she climbs up to the stigma of the flower and stuffs in the pollen ball, deliberately pollinating the flower. Few other pollinators do so intentionally; pollination is usually the unintentional result of a search for nectar. The pollinating behavior of the yucca moth ensures that the flower's ovary develops into a fruit, providing food and shelter for the moth egg and larva. As the larvae develop, they eat some, but not all, of the seeds in the fruit.

The plant and the moth's relationship is a remarkable example of mutual-

FIGURE 3-2

Yucca flowers, such as those of *Yuccu filamentosa*, are pollinated by the yucca moth (*Tegeticula yuccasella*), which gathers pollen from the stamens on which it is resting and then pushes the pollen ball into the central stigma of another flower.

ism. The moth larvae gain food and protection, while the plant is pollinated and produces seeds. The plant, however, loses some of its seeds to the hungry moth larvae, which feed on the seeds in the flower's developing ovary. Larvae consume between 6 and 25 seeds, but each ovary can produce up to 200 seeds. Studies indicate that about 13 percent of the yucca's seeds are lost each year to larvae of the yucca moth. Without the moth, however, seed production is dramatically reduced. (Although a few seeds may result from accidental pollination by bumblebees.)

When the seeds ripen, the moth larvae crawl from the fruit and drop to the ground, where they pupate. The adults emerge during the plant's next flowering season, but not all pupae emerge in any given year. Instead, each brood of pupae emerges over several seasons, ensuring that the species will survive even if there is a year when flowering or fruit-set fails.

## What Good Is a Mosquito?

We tend to think of organisms that bother us as unfit to exist, yet all have a role to play in the economy of nature. Most people consider mosquitoes, for instance, as nuisances the world would be better without. But would it? Adult mosquitoes are primarily active at night and serve as important sources of food for night-flying bats, nighthawks, and whippoorwills. Their aquatic larvae are an equally important food for species as diverse as young sunfish and predatory beetles. Mosquitoes also pollinate plants.

Mosquitoes are regarded as blood-suckers because the females require a blood meal before they are able to lay eggs. But mosquitoes also eat nectar, and male mosquitoes, in fact, only eat nectar. Nectar, while rich in sugar and calories, is deficient in protein, and it is protein that the females require for the development of their eggs. Female mosquitoes have solved this protein deficiency problem by resorting to drinking blood, which is loaded with protein. Male mosquitoes, however, don't need protein and are able to develop sufficient sperm with the nutrients they acquire from nectar alone. Mosquitoes, therefore, pollinate flowers as they search for energy-rich nectar, and some flowers depend on them for pollination.

The best-documented of mosquito-pollinated flowers is the blunt-leaved orchid (*Platanthera obtusata*), a terrestrial orchid whose distribution is circumboreal and does not occur in the Southern Appalachians. The green woodland orchid, the Appalachian twayblade, and the green adder's mouth, which have similarly small, greenish flowers and also grow in moist locations, might be pollinated by mosquitoes. One study of cranefly orchid identified a nocturnal moth as a rather ineffective pollinator; perhaps another nocturnal insect, such as a mosquito, is more effective. If mosquitoes disappeared, would these orchids follow them into oblivion? The question can't be answered definitively, but it is possible that their numbers would decline because pollination success decreased.

## Seeds and Fruits: Doll's Eyes and Hearts-a-Bustin'

By late summer and into early fall, many of the spring-blooming flowers have set seed. Some of these seeds, enclosed in fleshy fruits, attract animals to eat the fruit pulp and disperse the seeds, but few of their dispersers are known. It is a fertile area for study.

Doll's eyes (*Actaea pachypoda*) is one example of a fleshy fruit whose disperser is unknown. Each large, white berry has a conspicuous black center, is 1/2 inch in diameter, and does, indeed, resemble a doll's eye (Plate 23). Several berries are held aloft on a slender stalk six inches or more in height that arises from the upper part of the two-foot-tall plant. Two or three large compound leaves are showy even without the spectacular fruit. All parts of the plant are poisonous to humans and probably other mammals, which is why it also has the common name of white baneberry.

As with other plants that have poisonous fruit, dispersal of doll's eye seed is not well understood because fruits are normally consumed before the seeds are dispersed, which would poison the dispersing animal. In most fruits, the process of ripening reduces the levels of poisonous, or simply unpalatable, compounds, and a change in fruit color signals to visual animals that the fruits are ripe. In addition, mammals, birds, and insects respond differently to chemicals; what is poisonous to one group may not be to the other.

Hearts-a-bustin' or strawberry bush (*Euonymus americanus*) takes the prize for having the most brilliantly colored fruits and seeds. Before the fruits open to reveal the seeds, they look somewhat like pink strawberries, about 3/4 inch in diameter. Once the fruit ripens, however, the scarlet seeds inside burst through the wall of the pink fruit (Plate 24) to entice an animal to pluck them. Their dispersers are also unknown, but they are probably dispersed by birds, for their small, sparse leaves become colorful advertisements as soon as the fruits ripen, like those of other bird-dispersed plants. The plants reach several feet in height and are woody, with their attractive green stems providing some color during the winter months.

## Rain, Fungi, and Connections

Rain may make it difficult to enjoy hikes outdoors, but it is essential to the impressive biodiversity of the Southern Appalachians. Besides the abundant plants and animals that flourish because of these life-giving rains, fungi are equally diverse.

The mushrooms that pop up on lawns, trails, and tree stumps are just the reproductive part of a much larger, but unseen, mushroom body, the mycelium. A fungus is composed of threadlike strands of tissue that merge and tangle together in a network, but the individual threads, called hyphae, are so small that a magnifying glass is usually required to see them. The mycelium, however, can be huge (see chapter 4).

Some fungi are plant parasites, invading the living tissues of the host plant and causing its early demise. Many other fungi, however, are beneficial to plants. An estimated 90 percent of forest trees (as well as many other plants) have formed a symbiotic partnership with fungi called a mycorrhiza, which means "fungus-root." The hyphae envelop the plant root and sometimes penetrate into its cells. Fungi decompose materials by releasing enzymes from the hyphae into the soil and then absorbing the digested nutrients back into their bodies. Mycorrhizal fungi share some of these absorbed nutrients with their plant partners. In return for their service, the plant provides the fungus with some sugar produced by photosynthesis.

Some plants are obligately dependent on fungi for their growth. Notably, many orchids, such as the beautiful pink moccasin flower (see chapter 2), and some gentians (locally, *Obolaria virginica*, *Bartonia* spp.) are very difficult to transplant successfully because they absolutely depend on the success of their fungi. Not just the plant but also the better part of the fungal mycelium must be moved for a successful transfer, and this is rarely possible. Similarly, some of the clubmosses (see chapter 5) germinate and grow as small plants (in the gametophyte stage) only in the presence of mycorrhizae. Without the fungi, the plants cannot complete their life cycle. Germination of the tiny seeds of many orchids and heaths depends on fungal partners as well, for the seeds are so small that they carry few nutrients for the developing plant; the fungi supply those vital nutrients instead.

In a sense, mushrooms are the flowers and fruits of the fungus world, usually appearing for only a few days to release spores, and then disappearing until next season. The function of the mushroom is to hold the spores up where they can be dispersed to new locations by wind, water, or animals. Some of these mushrooms are even specialized, as are some plants, to attract animals that disperse their seedlike spores. In many fungi, those spore dispersers are flies and beetles.

Stinkhorns attract flies and carrion beetles with a strong smell that emanates from the spore mass, which is a slimy cap on the tip of the horn. The insects crawl around over the sticky surface, become covered in spores, and then fly off, rubbing off stinkhorn spores on whatever they touch next. Some insects as well as slugs eat the spores and then deposit them, unharmed, with a load of rich compost a few hours later. Most stinkhorns are also easy for humans to find because of their strong smell.

The elegant stinkhorn (*Mutinus elegans*) is a pink, conical mushroom shaped like a goat's horn, gently curved and tapering from base to tip. It is about six inches tall. It is usually enveloped with a white veil draped enchantingly around its middle portion, like a dancer's short skirt, and is topped by a slimy green mass of spores. Ravenel's stinkhorn (*Phallus ravenelii*) is more common but not nearly as elegant. These six-inch-tall mushrooms are capped with a pale head that is covered by the slimy green spore mass (Plate 25). Clusters of stinkhorns may appear along the wood-mulched paths of a manicured park or garden over a period of just a few hours. The horn hatches out of a ball of tissue that looks like an egg, and remnants of this white egg are often visible at the horn's base.

Other mushrooms rely on insects in a much more insidious way: they invade the body of a living beetle grub, feed on its tissues, and eventually kill it. When the beetle dies, the fungus matures and sends up its fruiting body above ground to release spores. The mushrooms of beetle cordyceps (*Ophiocordyceps melolonthae*) are orange stalks about an inch in height that appear in late summer and early fall. If you dig carefully at the base of the stalk, the mummified beetle larva or pupa is clearly visible (Plate 26).

Insects are not the only animals that disperse the spores of mushrooms. As I carry home clusters of the choice edible sulfur shelf (*Laetiporus sulphureus*), spores waft from my open basket into new locations. Easy to identify, sulfur shelf grows on dead trees in a stack of caps that are orange on top and yellow underneath when they first emerge (Plate 27). Instead of gills, this polypore fungus has tiny pores on the cap's undersurface from which the spores are released. Each cap can weigh a pound, and a cluster of fifty caps is not uncommon. As they age, they become white, flaky, fibrous, and tough.

Other tasty edible mushrooms belong to the chanterelle family. Cinnabar red chanterelles (*Cantharellus cinnabarinus*) are only an inch or two in height but are conspicuous because they are bright orange red. Like their delectable relatives black trumpets (*Craterellus cornucopioides*), they grow in groups, brightening the forest floor during the late summer. Both are good introductions to edible chanterelles because they have no dangerous look-alikes. Others, such as the highly desirable *Cantharellus cibarius*, require more careful scrutiny because dangerous species look similar. If you plan to partake, be absolutely certain of your identification by using a field guide to mushrooms.

While not edible, bird's nest fungi are both common and intriguing. These cute little nests are found grouped on dead wood and wood chips and, like stinkhorns, commonly appear in suburban areas, especially mulched flower beds. Splash cups (*Cyathus striatus*) are about 1/4 inch high, and within each little nest, several white "eggs" or masses of spores rest (Plate 28) until a drop of rain splashes them out.

Puffballs-in-aspic (*Calostoma cinnabarina*) are also very visible mushrooms, but they grow mostly in forests. When they first appear, the inch-wide red puffballs are embedded in a thick, clear jelly that really does resemble the gelatinous aspic glaze often used to decorate delicious hams and pâté. As they mature, tiny bright-red pieces break off like seeds. As the aspic eventually disappears, the mushroom dries out a little and the red color is restricted just to the tip of the puffball, where a raised ridge stands out like a set of pouting, painted lips. At this stage, they are often called pretty lips. Later in the fall and winter, the pale underground stalk of the puffball pushes out of the ground to a height of three or four inches.

Less cute but just as common are several species of slime mold. These masses of brightly colored slime (Plate 29) appear overnight, solidify, become firmer as they fruit, and then disappear in another day or two. It is easy to imagine that they are invading blobs from outer space—or worse, as the name dog vomit slime mold (*Fuligo septica*) implies. Slime molds rely on wet conditions and are most common in late summer after a series of soaking rains. Most grow on decaying wood, especially mulch, or grass. Slime molds are protists, but, like fungi, they spend most of their nonreproductive season underground and when conditions are right emerge to make and release their reproductive spores. Unlike fungi, they live not as a mycelium with strands of tissue ramifying throughout an area but as a group of separate amoebalike (protistan) cells. To reproduce, these cells all congregate into a plasmodium, which is what we notice above ground.

## Dangerous Social Insects and Their Mimics

The warmth of summertime swells the ranks of insects. By August, large and aggressive insects are hard to miss. Colonies of the social wasps and bees have reached their numeric peaks. Their "social" nature, however, is restricted to other members of their colony, to whom they are related as sisters. Within this insect society, there are strict divisions of work. There is only one queen, and she is the only reproductive member of the colony. She alone lays eggs, and the vast majority of these eggs develop into infertile female workers. The egg-laying ovipositor of worker females, which is unnecessary since they do not lay eggs, has been modified into a stinger. Instead of laying eggs, the sterile female warriors jab their stingers into prey or predators that attack their communal nest. A few drones, or males, are produced at the end of the summer solely for the purpose of mating with new queens, which are also produced in late summer. Lacking an ovipositor, males are unable to sting. In most native species only the newly mated queens overwinter; the rest of the colony dies after a few hard freezes.

Stinging insects are more visible in late summer for two reasons. For one thing, there are simply more of them. Their cycle begins in the spring, when mated queens awaken from their protected retreats under bark or leaves.

These queens build the initial nest, lay eggs, and raise the first brood. Once these wasps have grown into adults, the queen does nothing more except continue to lay eggs. As each brood of workers matures, it takes on different functions. Typically, the younger workers remain in the nest to care for larvae and the older workers forage afield for food. By late summer, numerous broods of workers have hatched.

Late summer is also the time of warmest environmental temperatures, another reason these insects are more visible. Insects, like reptiles, are warmed by heat from the environment. In all animals, bodily functions occur more quickly in warmer temperatures, producing faster reaction times. In the warmth of summer afternoons, insects are not only more active in their search for food but are able to respond quickly to disturbances near their nests. And it is a disturbance near the nest that is most often the cause of serious stings by these insects. Because they nest communally, a large contingent of wasps occupies the nest. They guard the nest, which contains larvae, just as a mother bear guards her cubs. If the nest is threatened, even unintentionally, the wasps respond with the only deterrent they have—their stings.

Ants, wasps, and bees are all considered social insects. Most wasps are carnivores that eat other insects and are therefore quite beneficial to gardeners. Most bees, such as honeybees (see chapter 2), bumblebees, and our native solitary bees, which most folks know as sweat bees, eat the nectar and pollen of flowers, serving as important, if accidental, pollinators in gardens and natural areas.

Ants are grouped with wasps and bees because they, too, form a complex society of different castes of individuals. Ants are also winged, or at least the reproductive queens and kings are, but the mated queens bite off their wings once the mating flight is over, forever remaining terrestrial in their underground nests. Ants are the most diverse of the social insects, serving as important pollinators, seed dispersers, scavengers, and even fungus gardeners. Some ants are able to sting, although few with such vigor as the wasps and bees, and most can bite ferociously (as can wasps). A few wasps also lose their wings but retain their potent stings; these "velvet ants" are hairy and reddish in color and pack a very painful punch.

Eastern yellow jackets (*Vespula maculifrons*) are the only wasps that commonly nest underground, and are the ones with which we most often come into uncomfortably close contact. It is easy to step on the opening to the nest before even noticing the small, clean hole in the ground from which a constant stream of jackets emanates. Nests of related bald-faced hornets (*Dolichovespula maculata*) look very similar to those of yellow jackets except

If you have wasp nests nearby, what can you do? If possible, leave them alone! The social wasps and bees sting only if the nest is threatened or if you step on or otherwise provoke them. Wasps are wonderful predators of pests in your garden, for they climb all over the leaves and are not picky about their insect prey. The Alabama farmer who had the huge yellow jacket nest probably had no worms on his tomatoes that year!

If you must destroy a nest (such as one in a doorway), crush it before it gets too large, and do it at night. All the wasps remain in the nest after dark. To get rid of an underground nest of yellow jackets, pour a gallon or two of very hot, soapy water over it at night. Again, approach the nest only after dark. The old-timey remedy of gasoline and the new-fangled pesticides are both toxic to life other than wasps, they poison groundwater, and their effects are long-lived. Avoid them.

---

they are hung on tree limbs. Both the nests of yellow jackets and hornets are covered with a gray papery material that sheds rainwater. The actual nest, made up of multiple layers of brood cells, is inside this covering. Nests begin small, about the size of a golf ball, but are constantly enlarged and may reach the size of a basketball or larger before winter kills the wasps. Paper wasps build nests that are a single layer of brood cells. Because they lack the outer paper covering, they are often built in protected locations under the eaves of buildings. Unlike honeybees, which make their brood cells from wax they secrete, wasps collect wood and chew it into a papery pulp; different colors in the nest reflect different sources of wood.

Sometimes unseasonably warm winter temperatures encourage whole colonies of yellow jackets and other social insects that are normally killed by the onset of winter to overwinter. In 2006, dozens of enormous yellow jacket nests were reported from Alabama. Some completely filled the interior of abandoned cars and were estimated to contain over 250,000 insects.

The distinctive marks on these social wasps advertise their dangerous sting. Yellow jackets are bright yellow with black rings and are usually less than an inch long. Bald-faced hornets are shiny black wasps with contrasting white rings and are about an inch long. Few animals dare to attack such well-armed prey, but some predators do. Some of the flycatchers, such as eastern kingbirds and eastern phoebes, capture individual foraging honeybees. Bears, skunks, and opossums will raid honeybee hives and dig up yellow jacket nests to dine on their larvae.

Some species of insects have also adopted warning colors even though

If your zucchini and yellow squash died last year just about the time they started to produce fruit, suspect squash borers. If you find small holes in the plant stems, but the plants are still healthy, ram a piece of wire into each hole to skewer the worm. If the plants have wilted, however, the worms have won. Prevention is the best medicine. The adult moths are present in late June, when they rest on the leaves or fly about to mate. Catch and crush any adults you find. Some gardeners cover their plants with row covers to foil the moths, but pollinators are also prevented from reaching the flowers, which prevents fruit production. Organic pesticides such as neem, pyrethrin, or rotenone kill the pollinators just as effectively as the pests. Restricting the chemicals to the stem base and applying pesticides only in the late evening after the blooms have closed will minimize the damage. Piling soil around the base of the plants, right up to the first true leaves, discourages but doesn't prevent the moths from laying eggs. A daily spritzing of soapy water on the squash stems is probably the best way to kill the eggs (as well as aphids on other plants) without harming pollinators. If plants are severely infested, pull them up and either burn them or crush them flat to kill the larvae; if you don't destroy the larvae, you will have more moths the following season.

---

they cannot sting and are perfectly palatable. Many species of flies, for example, mimic wasps. Hover flies (syrphid flies) mimic yellow jackets right down to the buzz. Not only are their markings bright yellow and black, but they are large and smooth-bodied. Some beetles, such as the locust borer beetle, have yellow and black markings to mimic stinging insects. The honeybee (Plate 30), another well-protected insect, is not as dramatically marked as a yellow jacket, yet it still has a recognizable form and behavior. The drone fly (*Eristalis tenax*), a member of the hover fly family, is such a good honeybee mimic that even biologists have a hard time telling it from a bee until closer inspection. This particular species (Plate 31), is actually a European species that was introduced into North America and is now common. The key difference between the fly and honeybee is in wing structure—flies, or dipterans, have only two wings, but wasps and bees, or hymenopteran, have four wings. When at rest flies hold each large wing out rather stiffly at an angle from the body, but wasps and bees fold their four wings neatly across the back. In addition, the abdomen of wasps and bees is separated from the thorax by a narrow waist, whereas flies maintain the same thickness of body through the thorax and abdomen. Flies also have small antennae made up of three segments, whereas wasps have long, ten-parted antennae. The eyes of flies usually occupy the entire head and touch each other on top, but the eyes of wasps are separated. Finally, flies have sucking mouthparts, but wasps have chewing mouthparts. These details are hard to see in a moving insect, making the mimicry extremely effective.

Several moths also mimic wasps, and one of these is the squash borer moth (*Melittia cucurbitae*). These bright orange and black diurnal moths are about an inch long and have transparent hind wings (Plate 32). They fly low and very fast. The clear wings, bright colors, size, and flight pattern all evolved to mimic dangerous wasps, thereby protecting the edible moth from predators.

Squash borer moths are serious garden pests. The females lay coppery eggs on the stems of squash plants just above the ground. The eggs, which are not easy to see, look like a row of tiny BBs on the green stems. The eggs hatch quickly, and the hungry caterpillars cut their way into the center of the stem. There the caterpillars live, safe from predators and from pesticides. They eat the heart of the squash stem, eventually killing the entire plant. A little pile of sawdustlike feces oozing from a hole in the stem is all that is visible.

## "Unsocial" Wasps?

Not all wasps are social and aggressive. The solitary species can be interesting to observe without fear of being stung. Most solitary wasps have bluish black bodies, some have colorful legs, and all flutter their wings as they hunt. Different species of wasp prefer different species of prey, including caterpillars, ground spiders, orb-weaving spiders, crickets, cicadas, and aphids. The battle between a wasp and its spider or cricket prey can be dramatic because they are of similar size; it is like a battle between a lion and wildebeest recast in miniature. The wasp leaps on its prey, and they tumble around together, rattling the dried leaves, as each tries to gain the advantage. Usually the wasp succeeds in stinging its prey, which then goes limp, but the prey may be so large that the wasp must drag it across the ground or fly in short hops to the nest. All these wasps feed on nectar as adults, but their young eat the insects that are brought to the nest.

The great golden digger wasp (*Sphex ichneumoneus*) eats crickets, and an individual wasp will often focus on a particular cricket species. The wasp digs a neat hole into sandy, hard-packed ground for its nesting burrow. The wasp always drags its cricket prey by the antennae and is unable to alter this instinctive program; if a researcher cuts off the antennae of the cricket, the wasp cannot drag the prey at all, even by a leg. The wasp always enters the

FIGURE 3-3
The nests of the organ pipe mud-dauber wasp (*Trypoxylon politum*) are cylinders of mud about four inches long. Inside each tube, several chambers are packed with paralyzed spiders to provide food for the developing wasp young.

burrow to check it before placing the prey inside, and if that same pesky researcher moves the cricket so that the wasp has to drag it back again, the wasp will check the burrow again, and again, and again with each move of the cricket.

The potter wasp (*Eumenes fraternus*), another solitary wasp, builds a little circular pot complete with a flared neck to hold its prey. I often find them attached to goldenrod stems. The small black wasp has creamy markings and fortifies its nest with caterpillars for its young.

The nest of the organ pipe mud-dauber (*Trypoxylon politum*) is characteristic of the species. The long cylinders of mud, six inches or so in length, are usually constructed in protected areas such as inside barns or under house eaves. Because several slender cylinders are constructed side by side, the completed nest looks like a rank of organ pipes (Figure 3-3). Although only used by the wasps for one summer, the nests may persist for years. The wasp builds the nest by collecting small balls of clayey mud in its legs and then molding each dollop of mud with its mouth and legs to the right shape. Often, it also vibrates its wings and buzzes, perhaps to dry the clay or even to help shape it. It has been suggested that the vibration of the wasp's wings is attuned to the resonant frequency of the mud, so that the mud picks up the vibration the same way a guitar string will hum with the sympathetic vibration from a correctly pitched voice. So much of nature is harmonic!

If you remove an active mud-dauber (or dirt-dauber) nest and look inside, you will find that the cylinders are made of several chambers, each of which is packed with spiders. Most people squeal, drop the nest, and take off at this point, but the story gets even more macabre. The spiders are alive but paralyzed because they have been stung by the wasp. Some of them will weakly move their legs, but none can crawl. And now for the worst of it: entombed with several helpless spiders inside each section of the nest is one wasp egg. When the wasp larva hatches, it will begin to eat the spiders, one by one. The spiders are kept alive so that they are fresh, ready for the larva to eat when the time comes. Similarly, sailors used to fill their sailing ships with Galapagos tortoises, turned upside down to keep them helpless, so that the crew would have fresh meat during their multiyear journeys.

Another group of strange-looking wasps feed their young on insect prey, but in an even more bizarre fashion. These wasps, called ichneumonids and pelecinids, have very long, skinny abdomens and ovipositors. Their prey is the larvae of beetles, and the long abdomen is used to penetrate either wood or soil, depending on where the beetle grub lives, and glue the wasp egg directly onto the body of the beetle larva. When the wasp larva hatches, it eats the beetle grub. The solitary adults, therefore, build no nest at all.

## Poisonous Butterflies and Their Mimics

The stinging wasps and their mimics form one large group of interconnected species. Poisonous butterflies form another. We generally don't think of butterflies as poisonous, yet several species are. It seems logical that large, showy, slow-flying butterflies must have some way of avoiding swift and active birds. While I've seen many birds (and one monkey) eat moths, I've never observed a bird eat a butterfly.

The best-known poisonous butterfly is probably the monarch (*Danaus plexippus*). It is poisonous because the larvae eat milkweeds, which themselves contain poisonous compounds, and the larvae sequester these compounds in their bodies and keep them into adulthood (see chapter 4). Another butterfly, the viceroy (*Limenitis archippus*), is a monarch mimic. It relies on birds' association of bad taste with boldly patterned orange and black butterflies, and it also tastes bad. As a larva, it feeds on willows and related plants and sequesters salicylic acid, the base ingredient of aspirin. Several other species are also likely mimics of the monarch. The common gulf fritillary (*Agraulis vanillae*), for instance, is also a boldly patterned orange and black butterfly.

A second group of butterfly mimics models itself on the poisonous pipevine swallowtail (*Battus philenor*). Like the monarch, it is protected by its poisonous larval food source. The larvae only eat plants in the birthwort (Aristolochiaceae) family, most commonly Dutchman's pipe (*Isotrema macrophyllum*). Wild ginger and heartleaf are also in this family of plants. These mostly black butterflies stand out when flying, resting on flowers, or feeding on nectar. Mimics of the pipevine swallowtail include the female spicebush swallowtail (*Papilio troilus*), the dark-form female eastern tiger swallowtail (*Papilio glaucus*), the dark-form female of the recently described Appalachian swallowtail (*Papilio appalachiensis*), the female eastern black swallowtail (*Papilio polyxenes*), the female Diana fritillary (*Speyeria diana*), and both the male and female red-spotted purples (*Limenitis arthemis astyanax*). All occur together in the Southern Appalachians, and birds probably do not distinguish between them.

Butterflies that are not poisonous are usually small and fly erratically, making it hard for predators to catch them. A good example is the eastern tailed blue (*Everes comyntas*). While it is difficult to get a good look at one, the effort is rewarded with the tiny butterfly's delicate beauty. Its wingspan, along with that of other blues, is about an inch, and its small antennae are black and white striped. The top of the wings are dark blue with a lighter blue border; from underneath, the wings' undersides are pale blue with dark specks. A threadlike tail extends off the hind wings. In males, there are two orange and black spots in the hind wings, but females lack these spots and are duller in color overall. The larvae are small, greenish, sluglike caterpillars that occur on legumes. Their favored food is clover.

The skippers also have an erratic flight pattern. A widespread skipper species is the silver-sided skipper (*Epargyreus clarus*). This common butterfly has a wingspan of about two inches. The wings and body are mainly brown, with two large white patches in the center of each hind wing and smaller orange areas on the forewings. The adults are frequently found on asters and other fall-blooming flowers, and the larvae eat the leaves of black locust as well as other legumes.

While butterflies might be encountered in fields or forests of flowers, they often concentrate around puddles, where large numbers flutter up when disturbed. Appropriately known as "puddling," this behavior is related to reproduction. The vast majority of puddlers are males, and they are drink-

FIGURE 3-4

The caterpillars of several butterflies escape notice by appearing to be something they are not. In the case of the viceroy caterpillar (*Limenitis archippus*), the caterpillar mimics the white, black, and greenish splash of bird dung!

ing from the puddles in order to increase their body's concentration of salt. When the males mate with a female, much of this salt is transferred to her, and she transfers it to the eggs she lays. Butterflies also gather around fertilizer and animal droppings because these sources also supply salt. Most plants are relatively low in sodium, yet it is a necessary nutrient. Herbivores often have trouble finding enough, as any cattle farmer who supplies a salt block for his herd knows. The butterflies are just gathering around their own salt lick.

Butterfly larvae are also often mimics. The larvae of many species of caterpillars try to escape notice by appearing to be something they are not; in the case of viceroy and red-spotted purple, they mimic bird dung (Figure 3-4)! Some of the large caterpillars, such as those of the tiger and spicebush swallowtails, mimic snakes. When disturbed, they protrude a Y-shaped horn called an osmeterium from behind their head, which looks like a snake tongue and also secretes stinky defensive compounds. Complete with eye-spots, they rear up and scare off small birds that may be attacking them (Plate 33). They respond similarly to an inquisitive human, making them easy to identify and interesting specimens on natural history tours! Inchworms in the family Geometridae (earth-measurer) look like small twigs. Their first reaction to a disturbance is to lift their front end up off the branch and freeze, thereby resembling a twig. Many treehoppers, insects that are related to cicadas and aphids, look like thorns.

## Moth Communication and Camouflage

Most butterflies, because they are diurnal, have good eyesight and respond to each other's visual cues and striking color patterns. Many moths, however, rely on odor, rather than vision, to find members of their own, primarily nocturnal, species. In moths, the chemosensory antennae are the dominant sense organs. As such, most moth antennae are typically featherlike, tapered from base to tip, with a large surface area.

Moth antennae will often distinguish males from females of the same species. Because the males use their antennae to locate a receptive female by following her scent trail, the antennae of males are larger and more feathery than those of females (Figure 3-5), providing a greater surface area for chemosensory receptors. Each antenna is covered with thousands of sensory hairs. When a male moth detects the chemical called a pheromone, which is specific to the female of his species, he flies upwind, following the intensifying odor trail, until he reaches her. The most impressive moth in terms of its capacity to detect female pheromone is the male silkworm moth, which can detect female pheromone in a ratio as small as one part pheromone in 80 quadrillion parts of air.

Polyphemus moths (*Antheraea polyphemus*) are an Appalachian relative of the Asian silkworms. During the day, polyphemus moths hang from the branches of trees like clusters of dead brown leaves. At night, they actively search for each other. Because the adults do not feed, they can concentrate solely on reproduction. The elaborate antennae of males distinguish them from females, and polyphemus moth males respond to females nearly as impressively as their silkworm moth cousins. One evening near dusk, I found a female polyphemus moth resting on the wall of a building. When I captured her, she immediately began to extrude eggs, indicating that she was either ready to mate or had recently mated. The next evening, in the exact location, I captured a male moth. He'd located precisely the spot where she had rested twenty-four hours previously.

Our chemosensory cells are found in the lining of our nose and mouth, but in insects, the sensory structures are located over the entire body. Insect mouthparts are covered with hairlike chemical detectors, but the legs and

FIGURE 3-5

Polyphemus moths (*Antheraea polyphemus*) have eyespots on each wing as a mechanism to startle would-be predators. The antennae of a male (inset) are larger and more feathery than the female's so that he can locate her by scent.

feet of many species are also densely furred. When a sugar solution is placed on their legs, flies, bees, and butterflies extend their mouthparts in order to feed. In fact, their legs are about 250 times more sensitive to sugar than a human tongue! The ovipositors of many insects are also covered with chemosensory structures, allowing the female to identify the potential food source for her larvae before she lays an egg on it. Research on butterflies has shown that females not only identify the plant species required by their larvae but also discriminate between individual plants of the species, choosing the one with the lowest levels of distasteful chemicals.

Moths are often described as furry because of the high density and elongated shape of their scales; the scales of butterflies are smaller. A major hazard for night-flying moths is spider webs. But the furry scales of moths detach easily, allowing them to slip out of many spider webs by sacrificing a few scales. In addition, the furry scales, like mammalian hair, trap heat, allowing moths to fly on cool nights.

Many nocturnal moths are sensitive to the ultrasonic squeaks of feeding bats. Some moths respond by folding their wings and dropping to the ground when they hear a bat approaching. A few moths, especially those in the tiger moth family (Arctiidae), even produce high-pitched squeaks themselves. Initially, researchers concluded that the moths were "jamming" the sonar system of bats by confusing it with false echoes. More recent studies, however, indicate that the squeaks may be acoustic warnings of distastefulness, serving a similar purpose as the bright color patterns of distasteful diurnal butterflies.

Moths must also protect themselves from diurnal predators. For example, polyphemus moths have a large, colorful eyespot located in the center of the upper surface of each wing. Normally, the moth rests with its eyespots concealed, but if disturbed, it responds by flashing the eyespots to startle a predator. Small birds, in particular, have been shown to back away when a moth flashes its eyespots. On such large moths (the wingspan of the polyphemus moth is over five inches), the eyes are sufficiently far apart to suggest a large predator to a small bird. Polyphemus is the one-eyed giant of Greek mythology.

Even the cocoons of these moths are camouflaged. They are brown, inch-long, leathery cases, usually incorporating leaves, wrapped in silken threads. During the winter, some fall prey to mice, whose sharp incisors can cut through the moths' silken retreats. How myriad are the connections in nature: you might hear the booming voice of a great horned owl one February night that has just eaten a mouse, which was kept alive because it found the overwintering pupa of a polyphemus moth.

## Luna Moths: Endangered or Not?

Luna moths (*Actias luna*) are in the same silkworm family (Saturniidae) as polyphemus moths. Like other adult silkworms, they do not feed and are short-lived. When at rest, these large, pale-green moths mimic leaves. The National Audubon Society's *Field Guide to North American Insects and Spiders*, by Milne and Milne (1980), reports that these moths are "now considered an endangered species because many have been killed by pollutants and pesticides." However, this statement appears to be unsupported by data. According to the U.S. Fish and Wildlife Service, the agency charged with monitoring threatened or endangered species, it is not listed as a federally endangered or threatened species. Larval food plants are common large trees such as birches, hickories (including pecan), black gum, sweet gum, persimmon, and walnuts, which are not exposed to most pesticides.

Anecdotal evidence, however, suggests that moths in general may be less

The true silkworm moth (*Bombyx mori*) is Asian in origin. It has been domesticated for thousands of years, artificially selected to produce the greatest quantity of silk. As a result, the adults have lost the ability to fly and are no longer found in the wild. Adult silkworm moths, like their close relatives, do not feed but simply mate and lay eggs. The larvae feed on mulberry leaves. As the larvae prepare to pupate, they spin a large cocoon with their mouths, and it is this cocoon from which our silk is derived. The cocoons are dropped into boiling water to kill the larvae and then painstakingly unwound into a single thread, which can be as long as 3,000 feet. About 2,500 cocoons are required to produce a pound of silk for a kimono; a silk shirt might only need 1,000 cocoons! Only a few pupae are allowed to emerge as adults, and they provide the eggs for the next generation of caterpillars. Silk has been produced in China since at least 2700 B.C., and that country continues to be the world's greatest exporter. All real silk is produced in this manner, dribbled from the mouths of domesticated caterpillars.

common than they were fifty or one hundred years ago. In the forward to the 1984 edition of Charles Covell's *Field Guide to the Moths of Eastern North America*, Roger Tory Petersen, respected observer of nature, describes an impoverished world of moths today compared to when he was a child. Of the same generation as Lorus and Margery Milne, perhaps he was witness to this general decline. Pesticide use may be a general factor, especially the massive chemical sprays for mosquitoes that are deployed over large areas of landscape by planes and trucks, but urban sprawl must also contribute. As we cover more of the Earth with asphalt, housing, and pest-free crops, nocturnal moths must cope with more distracting lights and a shrinking habitat.

## Katydids and Crickets:
## Summer Songsters Use Sound to Communicate

Some insects, such as grasshoppers, crickets, katydids, and cicadas, use sound to communicate. The dog days of summer hum with the buzz of the dog-day cicadas (see chapter 1), and summer nights thrum with the sound of chirping crickets. Many of the sounds of summer evenings are the calls of different species of insects as they strive to leave behind as many fertile eggs as possible, transferring their genes to the next generation of crickets. What we hear as peeps and chirps, crickets hear as the sounds of love, war, and hope for the future.

The insects rub the base of their wings together to make their calls. The inner edge of one wing, usually the left wing of katydids, but the right wing of crickets, is modified into a file, with teeth projecting downward. The other wing has the scraper, with teeth projecting up. Scraping the file produces the sound. When katydids sing, they rarely raise their wings, but field crickets, such as those sold for fish bait, hold their wings at a 45° angle, and tree crickets hold theirs at 90°. Grasshoppers rub a back leg against a wing instead. Only the males sing, for they are the ones who must communicate their desirability to the females and their invincibility to other males. Have a look next time you're in a store that sells crickets for bait, and you should be able to identify the boys that are creating all the noise by examining their wing posture.

The katydid is a leaf-shaped cricket. With a body that is about two inches long and legs and antennae that are longer than its body, this emerald teardrop is beautifully camouflaged among the green leaves of trees. The wings of katydids even have veins like those of leaves to disguise them further as they rest among the greenery. While there are many species of katydid, the common true katydid (*Pterophylla camellifolia*) is the only Appalachian species in which the hind wings are completely enclosed by the bulging forewings. Even its name reflects its affinity with leaves, for "ptero" means wing and "phylla" means leaf; "folia" also means leaf.

The body temperature of katydids and other cold-blooded insects matches that of the environment. The colder it gets, the slower they move, until, at last, they stop. As the nights get cooler, the orchestra begins earlier and plays more slowly, from "katy-did-katy-did-katy-did" to "kayy-teee-diddd." For me, it is this sound—the ever-slower drawl of katydids—that truly announces the arrival of autumn.

The snowy tree cricket (*Oecanthus fultoni*) is noted for its temperature-dependent call; the frequency of the call can be used to estimate the temperature in Fahrenheit within a degree or two. Count the number of calls it makes within thirteen seconds and add forty-one to determine temperature in degrees Fahrenheit. Its call sounds remarkably like that of a spring peeper, a tiny tree frog that is most vocal in early spring (see chapter 2). Like male peepers, male snowy tree crickets make a seemingly endless series of rapid, though musical, peeps. Both are calling to females of their species and announcing their presence to other males. If you think you hear spring peepers in early fall, they are likely to be snowy tree crickets instead.

Snowy tree crickets are widespread over the United States and abundant in the Appalachian region. They are pale green, delicate, and slender, about an inch long, with legs and antennae longer than the body. The males are paddle-shaped when viewed from above because their wings are wider than their bodies; they use those wings to produce sound for communication. Females have longer, thinner wings and bodies, and they hold the dark ovipositor underneath the abdomen. They use the short, strong ovipositor to jab eggs into stems of plants, and katydids slice into tree branches with their scythe-shaped ovipositor. Field crickets lay their eggs underground and have a longer, more flexible ovipositor (see Figure 1-3).

## Damsels and Dragons

While I was in the garden one summer afternoon, my hand brushed against something stiff and paperlike as I reached into my back pocket for some twine. Surprised, I twisted my head over my shoulder and saw a huge dragonfly perched on my hip pocket. A wasp was gripped tightly in its legs. The dragonfly manipulated the wasp, turning it so that the wasp's head was oriented toward its mouth, and then began to calmly crunch the antennae and head! My neck was stiff from observing the whole procedure, but the dragonfly remained on its unusual perch until it had consumed most of the wasp. Suddenly, it took off again as quickly as it had appeared and disappeared into the garden.

Dragonflies are such adept aerial predators that they are able to snatch their meals from midair, then land on a convenient perch to partake. All of them eat insects, including mosquitoes, flies and, as I witnessed, wasps. Some highly maneuverable dragonflies even eat other dragonflies. (At least one bird, the swallow-tailed kite, is a specialist on dragonflies too. No wonder the kites are such graceful and acrobatic fliers — they must be in order to catch those dragonflies!)

One of the most common dragonflies, which is also easy to identify, is the common whitetail (*Libellula lydia*). Only the males have a chalky white abdomen; females have a brown abdomen. Whitetails are active around shallow ponds, especially those without fish, where the males patrol the edges constantly. They watch for females or for intruding males. When a female appears, the male grabs her, they briefly mate, and he releases her as she begins to lay eggs. As she dips the tip of her abdomen again and again into the water to deposit eggs, he hovers over her, preventing other males from interfering.

The males use their white abdomens to signal each other. Each male establishes a small section of the pond's shoreline as his territory, but there is never enough shoreline for the number of males present, resulting in constant intrusions from rival males. When a resident male chases off a rival, he holds the tip of his abdomen upright, and the retreating male holds his abdomen angled downward. If you have a pond nearby, spend some time observing the different behaviors of males toward other males and toward females.

If you think the adults are dragonlike in their feeding habits, it is only because they learned them as offspring. The larvae are also voracious predators. They live underwater in streams and ponds, usually buried in the sediment or under leaves. When other invertebrates or even small minnows come by, they erupt from the bottom in a flurry of particles and grab them with large, pincerlike mouthparts that unfold from beneath their chin. They eat prey similar in size to themselves, just as do the adults.

Eventually, the larvae must leave the water and take to the air. They climb out onto sticks or leaves that extend above the surface of the water and then shed their larval skin and wait for their wings to unfurl (Plate 34). The molted skin looks somewhat like a small cicada skin. It takes some imagination to envision an adult dragonfly emerging from it. When I find a dried skin, I always stop to reflect on that transformation. From a life buried in mud, breathing in water, crawling along the bottom for months or years, a master of the air emerges in an instant, unchallenged except by the most agile birds. From nature springs philosophy!

Dragonflies and damselflies are closely related. Most damselflies hold their wings folded over their body when they are at rest, whereas dragonflies hold their wings outstretched. Damselflies are also small and slender, which is why they are considered damsels rather than dragons. Both are voracious predators and lay their eggs in water, where the young develop for months or even years.

A conspicuous Appalachian damselfly is the ebony jewelwing (*Calopteryx maculata*), one of the few damselflies that has black wings. The males are iridescent blue or green, truly jewel-like when illuminated by the sun. Females are duller in color and can be identified by white spots at the tips of their wings. The male jewelwings flitter and flutter up and down stream banks in a stylized and beautiful dance, resting only when neither competing males nor attractive females are present. They prefer small, slow-moving, shaded streams.

## Crab Spiders: Camouflage by the Predators

While prey species are camouflaged to protect them from predators, some predators are camouflaged from their prey. Crab spiders, an example of such an ambush predator that relies on its invisibility to capture its prey, vary in color according to the flower they usually inhabit. In our area, the golden-rod crab spider (*Misumena vatia*) is golden yellow and blends in so well with the flowers that they are hard to see until they have captured prey (Plate 35). They wait with four of their eight arms outspread in a welcoming gesture for any insect that visits the flowers. Large crab spiders may catch bumblebees, but smaller spiders capture flies and other insects.

## Flashing Fireflies of Summer Evenings

Throughout the Appalachians, summertime evenings are highlighted by animated stars, the dancing flash of fireflies. As darkness descends, flashes of fireflies rise higher and higher until by midnight they flash like Christmas lights in the treetops. Does every child scramble for a glass jar to catch that bit of flashing magic and then release it the next morning when the light of day transforms magic into skinny insects?

Fireflies or lightning bugs are neither flies nor bugs. They are, in fact, beetles. Like other beetles, their forewings are hard coverings called elytra, which protect their soft bodies and the papery hind wings that they use for flight. In order to fly, beetles cock open the elytra and extend the long flying wings. On the firefly, the lantern organ, which produces the light, is located on the underside of the abdomen near its tip, but different species have different numbers of segments, colors, and flash patterns. Males typically have more luminous segments than females, making their flashes brighter.

The largest local firefly is *Photuris pennsylvanicus*. At about 5/8 of an inch, it produces a green flash and is usually active well after dusk. (There is debate among specialists about the identification of the local species of *Photuris*.) The most common firefly in the Appalachians, as well as most of eastern North America, is *Photinus pyralis*. At about 1/2 inch in length, it produces a yellow flash beginning right at dusk. These two fireflies flash throughout much of the summer, but some of our most extraordinary Appalachian fireflies have a restricted season of engaging displays.

The synchronous firefly (*Photinus carolinus*), a close relative of *P. pyralis* that looks very similar, produces an eye-catching exhibition that occurs for about two weeks in mid-June. When the males flash, they do so in unison, all flashing together and all stopping at exactly the same moment, as if an invis-

FIGURE 3-6

It is easy to distinguish the female from the male blue ghost firefly (*Phausis reticulata*). The female (left) retains larval characteristics and never grows wings. The male (right) has wings and very large eyes so that it can locate the dim glow of the female. The male produces light from the two cream-colored light organs on the underside of the abdomen.

ible conductor has just tapped his baton on the podium. One firefly usually starts the chorus, then the rest join in, and sometimes an unnatural light such as car headlights can set them off. The trees, the shrubs, even the ground pulses with a steady throb of flashing yellow light; then it stops, and complete darkness settles in for a few seconds before they begin again. Elkmont, Tennessee, in the Smoky Mountains is famous for the fireflies' display, although they occur in other localities as well. I've seen them for many years in Balsam Grove, North Carolina, where they occur in a rich cove forest that rises up a steep hillside and has a small stream.

The blue ghost firefly (*Phausis reticulata*) is also common in the Southern Appalachians. Like the synchronous firefly, it is best known in one location (DuPont State Forest, North Carolina) but occurs throughout the region. The males are small, with black elytra and brownish wings, about 1/4 inch in length and delicate, but the females are flightless and grublike (Figure 3-6).

For visual animals, no nocturnal phenomenon is more engaging than a sudden flash of light. For us, this can be a searing meteor, a lightning strike, a fireworks display, or even a flashing neon sign. Light produced by living organisms, called bioluminescence, can be equally engaging. Bioluminescence is extremely efficient and produces essentially no heat, unlike incandescent light bulbs, which are only 5 percent efficient and lose 95 percent of the energy that flows into them as heat instead of light. Fluorescent bulbs are 20 percent efficient, which is why energy-conscious consumers use them. Not only do they consume less energy than incandescent bulbs for the same amount of light produced, but they also produce less heat. Bioluminescence, however, is 98 percent efficient, wasting only 2 percent as heat. Now we just need a bioluminescent lightbulb!

These small fireflies appear for about two weeks, their light display peaking around the first of June in fertile forest coves. They do not flash. Instead, they glow constantly with a dim bluish white light, drifting silently just inches off the ground. With hundreds or thousands of these fireflies meandering aglow over the dark forest floor, the ground itself seems eerily adrift. I could almost believe that the wee people were out, carrying small candles as they wander through the forest.

Fireflies use light to communicate. Both males and females flash, but it is the male who does most of the flying and most of the flashing. Each species has a unique flash pattern, a kind of Morse code, for the male signal and the female response. The males must use the right signal pattern and the female must respond appropriately or they will not mate with each other. A few firefly species, notably those west of the Rocky Mountains, are not bioluminescent at all but instead use pheromones to locate one another.

While adult fireflies are visible only for a short period of time each season, their larvae live much longer. The wingless larvae are predators in the leaf litter of the forest, where they eat earthworms, snails, slugs, and most anything else they can catch with their powerful jaws. They are the sharks of the lilliputian leaf-litter world, cruising for several months or years until they become large enough to pupate and transform into adults.

The larvae, as well as the eggs and pupae, of fireflies are also bioluminescent. Their constant, glowing light is means of advertising their distastefulness, for they are full of noxious compounds called lucibufagins. When the

Fireflies seem to be decreasing in number. Pesticide use, especially the unnecessarily large amounts used in suburban yards, probably contributes, but urban sprawl may be more to blame. Firefly larvae live in leaf litter for several years, relying on the stability of that system to provide them with food and shelter. When forests are cleared in order to build homes and roads or farms, firefly habitats are lost. Most firefly species pupate, emerge, mate, and lay eggs in the same small area they lived in as larvae. Other species are even more restricted because the females are wingless. In these species, only the crawling larvae can colonize new areas, and since they are only one-quarter of an inch long, they simply cannot move far over the course of the warm months of one year. Once eliminated from an area, therefore, they would be very slow to recolonize it.

A new program at the Boston Museum of Science is seeking to document the decreasing firefly population. If you're curious, join the firefly watch at <www.mos.org/fireflywatch>.

larvae, or glowworms, are disturbed, they increase their glow. Researchers have discovered that mice avoid glowing objects after one bad taste of a glowworm. In separate studies, a captive Swainson's thrush and jumping spiders rejected adult and larval fireflies as food. Bearded dragons, which are lizards native to Australia and presumably have no experience with fireflies, do not know to avoid them. There have been cases where pet bearded dragons have died after consuming several fireflies.

Noxious compounds notwithstanding, some fireflies may still become a meal. Female fireflies of the genus *Photuris*, for example, mimic the flash patterns of their smaller relatives, the *Photinus* fireflies, in order to capture and eat the males of that species. Lured by her bright, but counterfeit, flash pattern, the male *Photinus* flies to the alluring female *Photuris*, hoping to mate. Instead, she devours him, even sequestering his noxious compounds for her own use. These femme fatales occur throughout the Appalachians.

## Railroad Worms and Millipedes: Predators and Prey

In the Appalachians, larvae and wingless females in the beetle family Phengodidae are called railroad worms because each segmented worm resembles a passenger train with light emanating from the paired windows of its locomotive (head) and cars (segments). The "windows" are actually bioluminescent spots, larger on the head and smaller on the segments. It is likely, but

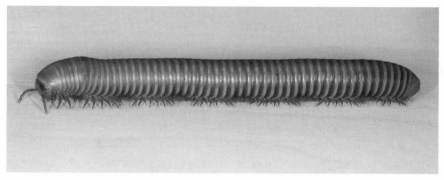

FIGURE 3-7

The four-inch-long, gray, cylindrical *Narceus americanus* is a very common millipede in the Southern Appalachians.

not confirmed, that the bioluminescence is an advertisement of distasteful-ness. Railroad worms feed almost exclusively on millipedes, and they may gain some noxious compounds from their food source, for most millipedes are distasteful.

Millipedes are repugnant because they release cyanide and other chemi-cals when disturbed. If you shake a millipede in your hand and then sniff it, it will smell of almonds, which can also produce cyanide (see chapter 4). Millipedes are slow-moving scavengers, and their only defense from preda-tors is their distastefulness; when disturbed, they usually curl up and wait for the noxious compounds to do the work. They are long, slender animals most often found in the leaf litter of mature forests. They resemble centi-pedes, but there the similarity ends. Millipedes have four legs on each seg-ment of their body, but centipedes only have two. Centipedes, as predators, have biting mouthparts, but millipedes do not. Centipedes often live under rocks or logs and scurry away when uncovered.

Millipedes and centipedes are archaic animals. A fossil of the millipede *Pneumodesmus newmani* was recently discovered. The millipede lived 428 mil-lion years ago in what is now Scotland and is considered the oldest land animal to have existed.

The Southern Appalachians have a huge diversity of millipedes, few of which are easy to identify except by specialists. The most common species of millipedes is the giant millipede, *Narceus americanus*. Its gray body is almost cylindrical and it reaches four inches in length (Figure 3-7). A common and pretty species is *Sigmoria stenogon*, which is flattened, red, about two inches long and 1/2 inch wide. *Boraria stricta* is black with yellow edges on each segment, also flattened, and about 1 1/2 inches long and 1/4 inch wide (Plate 36). It is one of the more potent producers of cyanide.

## Grouse Threat Display: Prey or Predator?

Sometimes the best defense is a good offense. When threatened, some animals that are large enough to do so respond by scaring a potential predator into retreat. For example, a garter snake once tried to convince me that it was really a rattlesnake, and I believed it, at least initially. Coiled up in a pile of leaves, it vibrated its tail to produce a sound remarkably like that of a buzzing rattlesnake. I jumped about two feet straight up and backed off, and it took the opportunity to quickly move off under some brush.

On another occasion, I was walking quietly along a forested path when, suddenly, coming rapidly down the path toward me, was a brown creature crouched low to the ground, making a high-pitched scream like I had never heard before. I jumped backward and shook the jacket I was holding in its direction, like a matador before a charging bull, imagining that it was a rabid raccoon. It abruptly turned aside and went up into a thick cover of ferns, all the while making this weird noise. It was no raccoon, I finally realized after several puzzling seconds; it was a ruffed grouse (*Bonasa umbellus*). Once I recognized it, I slowly backed up the trail and sat still. Soon, the bird quieted and started clucking gently. At this signal, a flock of chicks too numerous to count came out of the brush along the trail and climbed the bank toward her. Her charge, accompanied as it was by her extended wings, lowered head, and erect, fully-spread tail feathers, was an effective surprise that certainly backed me away from her chicks and almost ran me off. Against a smaller and less rational predator, the effect would be even more dramatic.

Similarly, I have observed eastern gray squirrels (Plate 37) charging against larger predators when their nest is threatened. Nimble and quick, they dash in for a quick nip against the flank of a black rat snake climbing the tree that supports their dray, which turns the snake around. While danger to the adult is real (a black rat snake can quickly throw a coil around an adult squirrel), the young still in the nest are completely unprotected against the snake.

The ruffed grouse (not ruffled) is named for its ruff, the group of black-tipped feathers around its neck. (Ruffs were deeply folded lace collars worn by European gentlemen in the sixteenth and seventeenth centuries.) A ruffed grouse is about half the size of a domestic chicken and spends most of its time on the ground. Well camouflaged, with brown feathers tipped with black or white, it springs into flight with a great burst of noise, which can be heart-stopping if it occurs right under your nose! It is the only grouse in the Appalachians, and in the southern mountains, we have reddish-phase birds with rufous tails. The male displays to females by strutting back and forth, often along a fallen log, with ruff, wings, and tail-fan spread out wide, beating his wings to produce a low "putt, putt, putt, putt, purrrrr" sound that is often compared to the sound of an old tractor starting up. Because the male appears as if he is beating his wings against the log on which he struts, like a carpenter hollowing out a tree trunk with a hammer and chisel, the Cherokee called it carpenter bird.

## Southern Appalachians: Greatest Salamander Diversity in the World

The abundant rainfall, high humidity, and clean water of the Appalachians are perfect for animals tied closely to water, and among terrestrial animals, the amphibians, or frogs and salamanders, are most dependent on it. They lay their shell-less eggs in water, and the gilled larvae remain there for their development. In addition, the scaleless skin of the adults must be kept moist so that it does not dry out, because they lack the cuticle of reptiles, birds, and mammals. Unlike our dry, rather flaky skin, the moist skin of the amphibian absorbs enough oxygen to supply most or even all of its metabolic needs. In the Southern Appalachians, the most common and diverse salamanders (family Plethodontidae) lack both lungs and gills and breathe directly through their skin.

For several reasons salamanders are more diverse in the Southern Appalachians than anywhere else in the world. In addition to the high humidity of the Appalachian cove forest, the rich leaf litter layer produced by the deciduous forest supports many small animals that serve as prey for salamanders. Because the Southern Appalachians are so old and have never been glaciated, there has been plenty of time for the slow-moving amphibians to diversify. One key geological feature, however, is primarily responsible for the large number of species. Rich coves, where most salamanders live, are often isolated from each other by high and dry mountain ridges that do not support salamanders. Only rarely does a salamander make it into a new cove. Physical barriers such as impassable ridges block interbreeding between isolated populations and allow them to diverge over time into new species.

The hellbender (*Cryptobranchus alleganiensis*) is the largest of North American salamanders at 2 1/2 feet long. In fact, the only larger salamanders in the world are its two close relatives that live in Asia. Hellbenders are impressive not only because of their size but also because of their appearance. Flattened to fit under rocks, with loose, floppy skin to increase surface area for absorbing oxygen, and colored to blend in with brown sand and dark rocks, they are grotesquely Mephistophelean. Not only do they look like they are from hell, where they are "bent" on returning, but they also "bend the hell out of" a fisherman's pole if they are hooked. As predators, they hide in wait to snatch small fish and invertebrates (including trout bait) that come near their den.

Hellbenders are becoming increasingly rare and are now mostly restricted to streams in the Southern Appalachians (isolated populations also occur in northwestern Pennsylvania). They require highly oxygenated, pristine rivers. Polluted water, introduced trout that eat hellbender eggs and young, and fishermen who catch and kill them take their toll on this species, which is long-lived and slow to reproduce. They mature at five years of age and can live for thirty. Like endangered pitcherplants, hellbenders are on the road to extinction because we damage their required habitat.

Perhaps our most visible salamander is the red eft, or eastern newt (*Notophthalmus viridescens*). Its life cycle is a bit unusual, and that is part of why it is so easy to see. Like other salamanders, the springtime eggs are laid in water, where they hatch into larvae. After a summer of feeding, the gilled aquatic larvae transform into the lunged, terrestrial efts. (The term "eft" is just an older English form of the term "newt." No other Appalachian species are newts.) The efts are juveniles, similar in size to the adults, but not reproductive. With a pair of lungs and thicker skin that affords more protection from desiccation, the efts can disperse far from their watery larval habitat. Depending on conditions, the eft stage can last from two to seven years. During that period, overland dispersal distributes the efts widely and provides ample opportunity for us to encounter them.

Eastern newts may be referred to as red efts, but Day-Glo orange is really a better description of their body color (Plate 38). Along their backs, there are tiny circles of even brighter orange encircled by black. This striking coloration is a warning to potential predators, for that pretty skin produces a potent neurotoxin. (I have observed dead trout with adult eastern newts in their mouths; presumably the skin secretions of the salamanders killed the trout.) The efts average three to five inches in length. They are strong and active, and look like tiny dragons or dinosaurs as they clamber over twigs and branches. Unlike the plethodonts, both the efts and adult stages are lunged. After their terrestrial sojourn is completed, the efts return to water, regain thin, moist skin, lose their brilliant coloration, and mature into adults.

The adults are aquatic, living in ponds, and in ponds on my property at least, they are the only common salamanders present year-round. The adults are about four inches in length, olive green on the back and dull yellow on the belly. They retain the efts' orange spots encircled by black, and these spots are the key to identifying them. In the spring and early summer, the adults undergo an elaborate courtship. They grasp one another and flip upside down, exposing their yellow bellies. Eventually, the male deposits a spermatophore and the female picks it up.

The adults and efts eat insects or other small animals they encounter. I've also observed the adults feeding on frog eggs. One recent spring, a horde of red eft adults completely consumed the first clump of wood frog eggs to appear in my ponds. A few days later dozens more clusters, containing hundreds or thousands of eggs, replaced the first clump, and I understood why the wood frogs lay so many eggs in such close proximity. The lone eggs were consumed, but the massed eggs overwhelmed the glutted adult efts and other predators. Many other natural phenomena, such as the masting of oaks and other nut trees (see chapter 1), function similarly to satiate their predators and ensure that some young of the prey species survive. As long as the balance between predator and prey is not disturbed, there will be enough individuals to support one another.

## Lizards and Snakes: Close Cousins

Only two lizards are commonly found in the Appalachian Mountains. The eastern fence lizard (*Sceloporus undulatus*) occurs from the mountains to the coastal plain. Males have a bright blue belly, which they expose to females by doing "push-ups." They prefer sunny, open locations, such as along wooden or stone fences, and are rarely found within the undisturbed forest. They reach about seven inches in length. Much of we know about the third eye has been learned by studying the fence lizard (see chapter 1).

Five-lined skinks (*Plestiodon fasciatus*), which prefer forested habitat, are more common. They usually have five light stripes on a dark body. Juvenile skinks have bright blue tails (Plate 39) that darken as they mature. While males of the broad-headed skink (*P. laticeps*) are the largest and reddest, all male skinks lose the stripes as they mature and develop a red head during breeding season. These large, red-headed skinks, contrary to popular myth, are not poisonous and are not scorpions.

Snakes and lizards are closely related, but several traits differentiate them from one another. While you might think that leglessness is the distinguishing feature, it is not a perfect one, for several species of lizards lack legs. Regretfully, the Appalachian Mountains rarely host the interesting legless lizards known as glass lizards, which are found in the coastal plain of the Southeast (although a few specimens have been collected on the western edge of the Great Smoky Mountains National Park). Fragile as glass, the tails of glass lizards are easily cast off, but, like those of other lizards, the tails regenerate. Snakes, who really do have tails extending beyond the cloacal vent, cannot regenerate them. (Tails of all vertebrates are composed of muscles surrounding the backbone and lack organs.) Leglessness, likewise, cannot be applied to all snakes. Some primitive groups of snakes, such as the pythons and boas, have retained traces of a pelvis and back legs. Two easily discernible features of lizards distinguish them from snakes: their movable eyelids, allowing them to blink, and external ear openings.

Most snakes have poor vision. Their eyes are permanently covered with a transparent scale that is clearly visible in the cast-off skin. During the process of shedding, as the old skin weakens, rubs off, and the new skin hardens, the snake is blind. Most snakes are also unable to change the shape of the eye's lens to focus on objects; they move the entire lens in a less effective manner. Snakes can detect movement, but most are unable to see sharp images. Lizards, by contrast, have excellent eyesight and good color vision, for they often use color in their displays. Male Carolina anoles (*Anolis carolinensis*) frequently erect their scarlet throat flap (dewlap) when courting females.

The ears of lizards are similar to those of mammals and birds. A small hole on each side of the head opens into an ear canal. Mammals, but not birds or reptiles, have an external flap of skin, or external ear, which surrounds the ear canal. The ear canal ends at the eardrum, which is a membrane that transduces sound from pressure waves in air to a mechanical vibration of the bones of the middle ear. In mammals, there are three bones of the middle ear (malleus, incus, stapes), but in reptiles there is only one (stapes). Embedded in the skull is the inner ear, which converts mechanical vibrations into impulses that nerves transmit to the brain.

Snakes lack external ears, ear canals, and eardrums, but they do have an inner ear and a stapes, which indicates that the ancestors of snakes likely had functional ears. Snakes can hear some sounds, but because they lack the eardrum as a transducer, their hearing is poor. They have been shown to detect low frequency sounds of 100–700 Hz and vibrations through the ground, such as a person walking, especially if their heads are resting on the ground, because their jaw bone transmits sounds to their skull.

The key characteristics of snakes—no legs, long and sinuous body form, no ears, and eyes protected by scales of the skin—are all thought to be related to the burrowing lifestyle. Presumably, the earliest ancestor of snakes burrowed through leaf litter and followed early mammals into their underground shelters. Front legs, in particular, could impede the animal's ability to slip down a narrow tunnel, and wiggling from side to side in a narrow burrow is more effective than moving with limbs. Ear canals would quickly plug up with dirt, and protective but clear covers over delicate eyes would be advantageous. Incidentally, open nostrils might also plug up with dirt, but forceful exhalations clear them. Most animals, including snakes, are able to use muscles to open or close their nostrils, allowing them to regulate their breathing.

The snakes' primary senses are taste and smell. Because a snake's nostrils are often closed while it is burrowing, continuous odor detection through the nasal cavity is compromised. So snakes detect odors swirling through the air by using their tongue. The flickering, Y-shaped tongue collects odor molecules from the air and carries them into the mouth. Each time the tongue is withdrawn into the mouth, it passes over a pair of chemosensory sacs, known as Jacobson's organ, on the roof of the mouth. The snake either drags the two tips of the tongue over the sacs or inserts them directly into the sacs, and the odors are detected there. Jacobson's organ is packed with sensory cells that are connected to the brain by a special branch of the olfactory nerve. Rather than potentially inhaling dirt into their nostrils and lungs as they track the scent of their prey by smell, snakes use their tongue to follow scent, dislodging dirt particles and cleaning their tongue each time it is withdrawn into their mouth. The two prongs of the forked tongue allow the snake to detect differing concentrations of scent, thus providing directional information, just as do two ears, two pit organs, or two eyes. A few lizards (monitors and gila monsters, for example) also have a Y-shaped tongue.

The snake is one of the most misunderstood of animals, perhaps because its characteristic features are not shared with other vertebrates. Although it is legless and slithers across the ground in what seems to be an absurd fashion, it can be disconcertingly quick. Then there is that cold, unblinking stare, which seems to mesmerize us mammals as we wonder whether or not the snake has even seen us. As we wonder, its strangely forked tongue flickers in and out, the only visible motion in what could be a carved likeness of a serpent. Lastly, some are poisonous and large enough to be dangerous to us.

## Snake Sense

Perhaps no other group of animals strikes fear into the hearts of so many people more than snakes. There is no need to fear every snake you see, though, particularly in the Appalachians. The copperhead and the timber rattler are the only poisonous snakes in the mountains, and you must nearly step on or otherwise provoke one to be bitten. Neither is aggressive. Neither will chase you. Water moccasins or cottonmouths are restricted to the coastal plain as are eastern diamondback rattlesnakes, pygmy rattlesnakes, and coral snakes. Countless harmless water snakes are killed every summer because they are assumed to be poisonous.

Timber rattlers (*Crotalus horridus*) are large, thick-bodied snakes up to six feet long and at least four inches in diameter. In the mountains, these snakes are often colorful, with yellow and black wavy bands and blotches. Outside the mountains, they are primarily brown or gray and also known as cane-brake rattlesnakes. In either locale, the last foot or so of the snake's body, excluding the rattle, is solid black.

Copperheads (*Agkistrodon contortrix*) grow to about four feet long and about two inches in diameter. They have a large, wide, coppery head. Their bodies are light brown with dark brown bands (Plate 40). Both copperheads and timber rattlers occur in forested areas and in fields that are bordered by forests, where they find their food source of mice, rats, and other rodents. Of the two, copperheads are more common. Contrary to the persistent myth, copperheads do not interbreed with nonpoisonous snakes.

Both the copperhead and timber rattler are pit vipers, which feed on small mammals. They are named for their specialized pit organs (Figure 3-8) that allow them to detect infrared energy, or heat, emitted by warm-blooded mammals and birds. The conspicuous pit organs, or pits, one located on either side of the head, are sensitive to energy wavelengths of 5,000–15,000 nm, and mammalian and bird bodies emit heat of about 10,000 nm. The pit organs can detect differences in heat over background radiation of only 0.003°C. Because the two pits are located on either side of the head but have overlapping sensory fields (like eyes) they have depth perception and are extremely accurate. A pit viper can strike as effectively in total darkness as in light.

Pit vipers use their poison to quickly subdue their mammalian prey. If you have ever tried to catch an escaped pet hamster, you probably know that cornered rodents can bite. Snakes face that potential danger every time they feed. Vipers strike quickly, retreat, and then wait patiently for the injected

FIGURE 3-8

Pit vipers, such as this timber rattlesnake (*Crotalus horridus*), lack external ear openings and eyelids but have conspicuous pit organs just below and in front of the eyes that are heat sensors. The nostrils in this snake, which are in front of and slightly above the wide-open pits, are barely open.

poison to immobilize their prey. If the snake acts fast enough, the prey does not run very far, and the snake saves energy in not having to track the prey over long distances. A snakes will use its venom defensively, but only as a last resort because the venom is costly to make and the snake must manufacture more once it has been expelled. In fact, human snakebites are frequently "dry," meaning the snake did not release venom.

Our common northern water snake (*Nerodia sipedon*) is often called a water moccasin and often mistaken for a copperhead because the two are similar in size and have similar markings, but the water snake is not poisonous. The northern water snake is brown with dark crossbands or blotches on its back (Plate 41). If completely dry, its back may be uniformly gray. The belly is lighter colored. The dark bands and blotches near the belly skin are red. As you can see in the photographs, the copperhead and northern water snake can be distinguished by the patterns of banding and background colors of the body, and especially the size and color of the head.

The snake that people are most likely to encounter in the Southern

FIGURE 3-9

The ring-neck snake (*Diadophis punctatus*) is a common, small, diurnal snake that can be identified by the yellow or orange neckband.

Appalachians is the black rat snake (*Elaphe obsoleta*). It is one of few snakes that is primarily diurnal, that is, active during the daylight hours. It is glossy black with a white underbelly. These snakes can climb trees, so you may encounter their shed skins dangling from tree trunks or heavy limbs. Their belly is flattened rather than round, which assists in this tree-climbing ability. Black rat snakes do, indeed, eat rats, in addition to other small mammals, so they are good to have around a farm and anywhere else rodents are a problem. They kill their prey by quickly coiling around it; but rather than slowly suffocating the prey, in many cases their squeeze elevates its blood pressure to the point of arterial burst. The biggest black rat snakes that I have ever seen were on a local farm, where I presume they had been eating well-fed rats. Although I did not harass these impressive specimens to straighten them out, they were at least seven feet long and four or five inches in diameter, which approaches the maximum reported size for the species!

If you want to get more comfortable around snakes, then the ring-necked snake (*Diadophis punctatus*) is the place to start. These small, elegant creatures are easy to identify by their pretty orange or yellow collars (Figure 3-9). Most are less than ten inches long and 1/2 inch wide, although they can reach a maximum length of twenty inches. They occur in forested habitats and frequently come to piles of gravel or mulch, where they probably lay their eggs. Their main food sources are earthworms and small salamanders.

## The Centenarian Turtles

Appalachian turtles are not as diverse as those found on the coastal plain. The most commonly encountered species are the mainly terrestrial box turtle and the mainly aquatic common snapping turtle. Sliders and cooters, which are widespread in the lakes and slow rivers of the coastal plain and piedmont, are represented in the mountains by only one native species, the painted turtle (*Chrysemys picta*), although other species have been introduced. The rare wood turtle (*Glyptemys insculpta*), like the wood frog, is a northern species that can be found as far south as the Southern Appalachians. The endangered bog turtle (*G. muhlenbergi*) occurs in mountain bogs and is losing its essential habitat.

Snapping turtles (*Chelydra serpentina*) are by far the largest local turtle, weighing as much as sixty pounds with shells measuring up to eighteen inches long. They are most often observed in ponds, where they eat invertebrates, small fish, amphibians, and plants and scavenge dead animals. They wander on land after springtime rains searching for mates and egg-laying sites.

Snapping turtles are known for their quick and easily provoked bites. Although turtles don't have teeth, their jaws are as sharp and hard as a parrot's bill, and a big turtle can inflict a serious wound to a finger or toe. A southern legend says that once a snapping turtle bites down, it will hold on until it hears thunder.

The aggressive nature of the snapper is undoubtedly related to its reduced shell. The upper shell, called the carapace, is smooth, relatively thin, and even transparent in dead specimens. The lower shell, called a plastron, is shaped like a small cross, as compared to the large, solid plate typical of other turtles. The turtle cannot pull its massive head, long tail, and feet into the shell to protect itself. Instead, its behavior protects those vulnerable parts from most animals other than humans, who favor it for turtle soups and stews. These turtles aggressively defend themselves on land, but they rarely bite if under water, where they are presumably safer.

Alligator snapping turtles (*Macrochelys temminckii*) are larger than common snappers and occur in the Mississippi River drainage; while they can occur in the Southern Appalachians, they are more often found farther south. Raised plates on their carapace and ridges on their tail distinguish them from the common snapper. Alligator snappers are predators that use the tip of their modified tongue as a lure to attract small fish, but enticing a snapper to open its mouth to get a look at its lure is problematic at best!

Box turtles (*Terrapene carolina*), our other common turtle, are more terrestrial. They occur up and down the eastern half of the country and live in many different forested habitats, from the mountains to the coast. During periods of dry weather, they often retreat to ponds or streams. These turtles spend the winter buried underground at a depth of two feet. They choose an area with loose soil that they can easily dig into and use this general location each year for their wintering ground.

Turtles have an average life span of seventy years, and they occasionally live for more than one hundred. They do not reproduce until they are five years old. I'll move a turtle that I find on the road because the possibility that it has been around for ninety years without being killed by a natural predator makes it precious indeed. I only move the turtle off the road, as opposed to taking it home, because these turtles can be hard to introduce into new areas. The adults have distinct home ranges with specific trails, sources of water, and wintering grounds. These home ranges are usually no more than 250 yards in diameter. If they are moved into new habitats, many will die over the winter because they are unable to locate those resources before winter sets in, or the new territory may already be occupied. In general, it is best for them to remain in the same area in which they hatched.

Most male and female reptiles can be difficult to distinguish, but that is not the case for box turtles. Most adult males have red eyes, and all females have brown or yellow eyes. Immature males and occasional adult males have brown or yellow eyes, too. The shape of the turtle's shell, though, reveals its sex for certain. The upper, dome-shaped carapace is attached to the lower plastron, which covers the belly. In females, the plastron is perfectly flat, but in males the plastron is cupped, or concave, so that the male can better balance on the female when they mate.

The plastron of the box turtle is hinged in both male and female so that the turtle can tightly close its shell to protect its head, tail, and legs. The turtle is a rather docile animal, probably because it is so well protected. Few adult turtles can be killed by predators, but the young ones are eaten by many different animals, including raccoons, several kinds of birds, dogs, and cats. Box turtles eat a variety of plants and fruits, in addition to insects, earthworms, slugs, and dead animals, but their favorite food is mushrooms. They even eat mushrooms that are poisonous to humans, and some people have been poisoned by eating a turtle that had eaten a poisonous mushroom.

## Climatic Conditions in Summer

The Appalachian summertime is rich with flowering plants and the insects that pollinate them. Abundant rainfall encourages impressive growth of plants. The warm temperatures accelerate plant growth and animate cold-blooded animals, such as insects, amphibians, and reptiles, which thrive in the tropical splendor, along with the warm-blooded birds and mammals.

In the Great Smoky Mountains in Gatlinburg, Tennessee, in June, the average high temperature is 82°F and the average low is 57°F. The longest day of the year, the summer solstice, occurs on June 21st or 22nd, when the sun rises at 6:18 A.M. and sets at 8:53 P.M. At about 14 1/2 hours, June has the longest average day length of the year. June is one of the wettest months of the year, with an average of 5.8 inches of rain over ten days.

In July, the average high temperature is 85°F and the average low is 60°F. July is not only the warmest but also the wettest month of the year, with an average rainfall of 6.0 inches over ten days. Average day length in July is about 14 hours.

The average high temperature in August is 84°F, and the average low is 60°F, with an average of 4.6 inches of rainfall over eight days. The average day length in August is 13 1/2 hours.

# Cycles of Fall

## September, October, November

4

Fall is a season of restlessness, a time of change from the near-tropical lush-
ness of summer to the cathedral-like solemnity of the upcoming winter.
The cool fall winds provide welcome relief from the heat of summer, help
transport migrating birds and insects, and animate the landscape. A rush
of colorful butterfly flowers and butterflies precedes the breathtaking transition of the verdant forest into a sea of brilliant color. As they flutter to
the ground, the innumerable leaves seem to mimic the exuberant color and
motion of the incomparable butterflies. The first frosts not only intensify
color and activity but also promise a respite from the burgeoning numbers
of insects. As winter nears, a few dragonflies, bumblebees, and pesky yellow
jackets linger, slowing down day by day before finally disappearing until
next year. The late-flowering gentians, as blue as the cerulean sky, seem re-
luctant to open their tightly clasped blossoms against the chill. The closing
act is played by witch hazel, whose wild, twisted, yellow blossoms openly
defy, yet eventually succumb to, the onset of winter.

Fall color in the Appalachians is a legendary attraction. The delightfully
cool air pours down the mountainsides, scented with the aromas of leaves,
ripe fruit, and late-blooming flowers, like the bouquet of fine claret. The
trees are awash in color. The deep yellow of hickories competes with the
sun for brilliance, while maples clad in shades of red, yellow, and orange
paint the landscape. The purple leaves of sourwood as well as the yellow
of sassafras and Fraser's magnolia seem to glow by some inner light. The
immense forest is a canvas of plaids, and fallen leaves swirl in multicolored drifts. On a finer scale, cinnabar red chanterelles decorate the forest
floor beside black trumpets, pink elegant stinkhorns, purple club coral, and

orange earth tongues, a few examples of the colorful fungi that complement the leafy gallery overhead.

## Bioluminescent Mushrooms

Jack-o'-lantern mushrooms (*Omphalotus olearius*) erupt from the bases of living trees, usually oaks, during the autumn season. As orange as pumpkins and present during the season of harvest, they also emit an eerie glow, reminiscent of the spooky light of a candle in a carved pumpkin head. To see this dim, ghostly light, total darkness is required. Even moonlight will obscure the faint, haunting, greenish glow.

Not only does the mushroom itself glow, but so does the entire fungal body. A mushroom, which releases spores, is only a small part of the entire fungus, just as a fruit, which releases seeds, is only a small part of the whole plant. Most of the fungus is made up of tiny rootlike strands that are buried in the soil, and the mat of these strands is called the mycelium (see chapter 3). The mycelium grows throughout the ground and into the wood of the tree, and it also glows. Fragments of decaying, glowing wood with the fungus embedded in it are called touchwood, and this ghostly light is one of the types of foxfire (another is from atmospheric electrical energy). Both the ancient Greeks and American Indians described glowing wood in dark, nighttime forests, and World War I soldiers added touchwood to their helmets to prevent collisions with each other in the pitch-black trenches. The word "foxfire" is a corruption of "faux fire," or false fire.

Although jack-o'-lantern mushrooms are poisonous if eaten, they are not harmful to handle. They are sometimes confused with chanterelles because both are large and orange. Jack-o'-lanterns grow at the base of live trees, but chanterelles grow on the ground. While chanterelles are delicious, jack-o'-lanterns cause gastric upset for several days if they are swallowed. While they probably will not kill you, you may wish you were dead!

Another Appalachian fall mushroom is also bioluminescent. Honey mushrooms (*Armillaria mellea*) are considered a delicacy by mushroom gatherers, but they can be confused with poisonous jack-o'-lanterns as well as several other poisonous mushrooms. Like jack-o'-lanterns, honey mushrooms can be detrimental to living trees, because they do not necessarily wait for their host to die before infecting it. This same genus is famous for its extremely large mycelia. The largest organism on Earth, confirmed by genetic testing in 2001, is a fungus (*A. ostoyae*), whose mycelium covers 2,200 acres (3.4 square miles) of Malheur National Forest in the Blue Mountains of eastern

Oregon and is estimated to be between 2,400 and 7,200 years old. It displaced the thirty-seven-acre original "humongous fungus" (*A. bulbosa*) near Crystal Falls, Michigan. All the mushroom clusters in those areas of forest are genetically identical and part of the same individual mycelium. The function of bioluminescence in fungi is unknown, although it may attract insects that disperse the spores. The minute spores of most mushrooms are readily dispersed without the help of insects, but perhaps there is a particular species of insect attracted to the bioluminescence that helps the fungal spore invade living tissue. Nature still holds secrets as yet unlocked by science!

Now that few of us live in areas that are truly dark at night and even fewer of us are outside after dark, most people have never been bewitched by bioluminescence in mushrooms and other organisms. The dim light is difficult to photograph or record and might leave you wondering whether the mushroom cast a spell on your senses, but if you are willing to go outdoors after darkness falls, you may be transported by the subtle beauty of the natural world.

## Leaf Color Change

Red, orange, yellow, green, blue, indigo, violet: these are the colors of sunlight shot through a prism or the raindrops of rainbows. Together, they constitute the visible spectrum of light from longest (red) to shortest (violet) wavelengths. These are the wavelengths of energy that our eyes are attuned to see, although other animals can see additional wavelengths. Bees, for instance, can see the very short wavelengths of ultraviolet but sacrifice the red end of the spectrum to do so. Snakes, specifically pit vipers such as rattlesnakes and copperheads, can sense the longer wavelengths of infrared energy, also known as heat. Although they do not see actual images in infrared light, the pits have overlapping fields like those of eyes, allowing them to pinpoint the source of heat and accurately strike a target, such as a warm-blooded mouse (see chapter 3).

Leaves are green because they contain chlorophyll, which reflects green light while absorbing red and blue. We see the color of light that is reflected, not absorbed, by the leaf. By reflecting green light, however, the plant is unable to capture the valuable energy in that particular wavelength. It therefore traps as much as possible with additional, accessory pigments in the leaves, which capture some of the energy that chlorophyll misses and then transfer that energy to chlorophyll.

Chlorophyll is one of the most important molecules on Earth because it converts the energy of sunlight into the energy of organic molecules and transforms carbon dioxide into carbohydrates (see chapter 2), but it is an expensive molecule to build. A large part of the energy used by the plant is needed simply to maintain chlorophyll in ample supply. Chlorophyll also breaks down easily, especially when exposed to cold temperatures and, ironically, sunlight, which degrades chlorophyll in much the same way that it oxidizes housepaint or burns human skin. Another function of the accessory pigments, in fact, is to act as a sunscreen and protect chlorophyll from damage by intense light and oxygen. These "phytochemicals" (plant chemicals) are among the antioxidants that are promoted today for human health. Even with the protection of accessory pigments, a plant must constantly make new chlorophyll to replace that which is destroyed.

During spring and summer, the green of chlorophyll masks the colors of the accessory pigments, which include carotenoids and anthocyanin. The carotenoids are a group of closely related, common pigments that are present in most leaves (and carrots). They absorb green and blue light and reflect orange and yellow. Anthocyanin also absorbs green and blue, but it reflects red. Thus, the carotenoids appear yellow or orange and anthocyanin appears red or purple.

On a fall day, the hills are alive with these vibrant colors. So many leaves become various shades of yellow and red because the chlorophyll in those leaves breaks down, but the accessory pigments remain or are synthesized. As the days shorten, deciduous plants decrease the production of chlorophyll, and it quickly disappears from their leaves. Since the accessory pigments are more stable, they persist and delight our eyes. In some cases, plants even produce more accessory pigments, especially anthocyanin. Once the green of chlorophyll is gone, these other colors shine through (Plate 42).

Beautiful color appears every year at about the same time because the shortening days of fall signal the trees to begin the process of chlorophyll loss, but some years are more colorful than others. A warm and wet summer ensures that every leaf is packed with pigments and every tree is loaded with leaves, setting the stage for a spectacular show. During the peak period of leaf change, which is usually mid-October for the Southern Appalachians, several converging conditions produce a short, but extremely colorful season. A cold snap intensifies the color because cold temperatures break down chlorophyll quickly, thereby revealing the accessory pigments. Dry weather also intensifies the color because anthocyanin, present in sap, becomes more concentrated as the water in sap evaporates away. Excess water dilutes the sugary sap, as anyone who has tasted a watery, wet-weather cantaloupe knows. Finally, since sunlight also destroys chlorophyll, sunny weather intensifies the color. The best color show, therefore, occurs during a fall season that is dry, sunny, and cool at night. When these conditions are met, a short but spectacular show results, and it is the timing of the cold snap that determines when the show begins.

The variability of red in leaves, which ranges from bright pink to nearly purple, is related to pH. Trees with more acidic sap, such as red maple (see chapter 2), produce bright red leaves, whereas trees with less acidic sap, such as sourwood (see chapter 3), produce purple tones. Individual trees, especially of red maple, that are growing in soil of different acidity levels may have different color tones than their neighbors. You can even roughly estimate the level of acidity of a forest cove based on its predominant fall leaf color. If the leaves are mostly yellow, from sweet birch and other species, the soil of the cove is likely less acidic than would be the case if the leaves were mostly red.

Because Appalachian microclimates, such as those created by the difference in temperature between a valley or a ridge, can vary significantly, at least one locality is likely to have met these optimal weather conditions during the critical period of leaf color change. For example, north-facing and south-facing slopes may differ from each other in temperature, moisture, and plant species, and one may produce stunningly beautiful color while the other produces less remarkable color or produces it later in the season.

The great diversity of trees in the Appalachian region is another reason for predictably good fall color. With more than one hundred species of trees present, the southern mountains will feature beautiful leaf color in at least some species nearly every year. By contrast, the Northern Appalachians rely heavily on single spectacular species, such as sugar maple.

## Fall Flowers: Gentians and Orchids

While fading leaves, rather than flourishing flowers, provide most of autumn's color, a few floral surprises are still in store. Gentians are among these gems of fall, blooming mainly in September and October. Some local yellow-flowered gentians (*Obolaria virginica*; *Bartonia* spp.) depend on their mycorrhizal (see chapter 3) association with fungi to grow. Most gentians, however, have blue or purple flowers that barely open until forced open by a pollinating insect. Stiff gentian (*Gentianella quinquefolia*) is one of the species that grows along the open roadsides of the Blue Ridge Parkway. The entire plant is about six inches high, and the individual flowers are about an inch long. Clusters of flowers (Plate 43) emerge from the tip of the plant as well as from other leaf axils. It is unusual among gentians because it is an annual or biennial rather than a perennial.

Nodding ladies' tresses orchids (*Spiranthes cernua*) are also small but characteristic fall flowers, sometimes known as autumn ladies' tresses. The slender flower stalk is a foot or so in height, and it bears about twenty small white flowers that spiral around it (Plate 44). In this species, the base of each flower angles downward, like a little nodding head. They grow in locations similar to those of gentians, preferring open areas where they don't have to compete for sunlight. I see them most frequently along roadsides, where they often form dense colonies, especially in moist, mossy sites. Surprisingly hardy, they bloom until November, well after first frost but before hard freezes set in.

## Witch Hazel's Bewitching Flowers

Along wooded paths or stream banks in late October or early November, the unexpected smell of spring flowers has more than once stopped me in my tracks. It always comes as a surprise when the bewitching flowers of witch hazel (*Hamamelis virginiana*) begin to bloom.

Witch hazel is a shrub, rarely more than fifteen feet in height. Its gray branches twist and turn against the backdrop of surrounding plants or sky. It is most common along stream banks or in coves but is sometimes found

FIGURE 4-1

Witch hazel (*Hamamelis virginiana*) blooms in fall and winter. The peaked galls on summer leaves, which look like witches' hats, are caused by aphids (*Hormaphis hamamelidis*).

along the edges of fields or other sunny spots. Its roundish leaves are fuzzy to the touch and usually drop off before the plant blooms. Witch hazel derives part of its name from the fact that its leaves look like those of the hazelnut.

The flowers of witch hazel mock the standards of shapely, pretty flowers. Four long, narrow petals of yellow twist in sympathy with the tree's somewhat contorted trunks and limbs. Insects urgently search for a store of nectar and pollen from these last flowers of the year, effectively pollinating the plant. The tree produces an abundance of round nuts that release their numerous seeds the following fall when the next bloom graces the plant. The plant is also referred to as a snapping hazelnut because when they are ripe, the nuts burst open and shoot the seeds away from the parent plant.

The name witch hazel comes from several sources. The slender branches of witch hazel are favored by dowsers to find underground water, a practice also known as water witching. The strange blooming time, near Halloween, also makes the tree seem bewitched. In addition, insect galls that look like peaked witches' hats (Figure 4-1) cover the leaves in summer.

The galls, which are abnormal growths on the leaves, are caused by aphids (*Hormaphis hamamelidis*). As the aphid pokes holes in the leaf to feed, it injects chemicals that cause the leaf to respond with excessive growth. Eventually, the aphid is completely enclosed by the walls of the gall and then produces up to seventy offspring, which climb out of the gall, move to other leaves, and make more galls. These offspring are clones of the first, produced from eggs that were never fertilized. Only under stressful conditions, such as the approach of winter, are males produced, and then sexual reproduction takes place. The advantage to sexual reproduction is that it leads to new combinations of genes that result in aphids that differ from each other in various traits. After overwintering as eggs, these new aphids are then confronted by the circumstances of a new spring season. Some flourish while others perish. Those that flourish do so because they are best adapted to the new environmental conditions. Those select few then reproduce clonally, rapidly replicating themselves and their well-adapted genes to exploit the seasonally available resource.

Witch hazel is used medicinally in several ways. Because its bark is astringent, an extract was popular at one time as an aftershave lotion and cosmetic. It is still sold commercially as an aftershave and a balm for sore muscles, and it works! After an afternoon of gathering, splitting, and stacking firewood, the delightful coolness of witch hazel extract rubbed into tired muscles soothes away the ache as the familiar scent conjures up the delicate flowers of fall. Native Americans used it as a poultice for aching muscles and insect bites. Cultivars of native and Asian varieties of witch hazel are popular as shrubs in the horticultural trade. Identifying these numerous cultivars can be quite confusing; they can have red, pink, or yellow flowers, but they are usually shrubs (as opposed to small trees) and have small round leaves.

Common in every state east of the Mississippi, witch hazel is particularly abundant in New England and along the Appalachians. It was a familiar friend to Thoreau, who writes in *Autumn*, "While its leaves fall, its blossoms spring. The autumn, then, is indeed a spring. . . . By their color as well as fragrance, they belong to the saffron dawn of the year, suggesting amid all those signs of autumn, falling leaves, and frost, that the life of nature by which she eternally flourishes is untouched."

## Dispersal of Offspring in Plants

Fall is harvest time for humans and animals alike. While we juice apples, crack walnuts, pick pumpkins, make jelly, and can beans, animals are also busily gathering nature's bounty. Lacking the ability to store food in cans or freezers, however, they mostly store it in their own bodies, as fat. The fruits of fall are especially important in the diet of most animals. Botanically

FIGURE 4-2
Milkweeds, such as poke, or tall, milkweed (*Asclepias exaltata*), have wind-dispersed seeds with cottony fibers.

speaking, "fruit" is a very broad term, encompassing more than we generally envision, because a fruit is the product of a pollinated flower, or a plant's mature ovary, and therefore houses the seeds. Fruits are not only the fleshy grapes, raspberries, or plums that we relish but also the dry and woody or papery milkweed pods, maple samaras, and oak acorns that animals devour.

The function of a fruit is to protect and disperse the seed, which is the plant's offspring. Plants cannot move, and so their seed, like their pollen, must be moved for them, usually by wind or animals. Some fruits are designed for passive or accidental transport, with structural modifications that enhance their ability to catch in fur or travel on the wind. Many familiar plants, such as dandelion, willows, and milkweeds, have small seeds with tufts of gossamer that are dispersed by wind (Figure 4-2). The fruits of tulip trees and maples have little wings that twirl like the blades of a helicopter as they are blown from the parent tree. Other fruits, such as stick-tights, beggar's ticks, and cocklebur are covered with Velcro-like hooks that catch in the fur or feathers of animals.

Other fruits are designed to actively attract animals to disperse their seeds. These fleshy fruits are edible fruits that enclose the seeds. The seeds are dispersed when animals feed on the fruits, either by swallowing the seed or discarding it after they have carried it somewhere else. Seeds that are swallowed must be small and protected so as to survive passage through a gut, but will be deposited with a nice load of fertilizer when the animal defecates. The germination of some seeds is actually enhanced by exposure to acid, such as that in a stomach, because it weakens the very tough seed coat.

Some seeds, however, are not designed to be eaten even if the flesh of the fruit is. The seed of a peach, for instance, not only is enclosed in a very hard seed coat, but it also contains a form of cyanide to make it inedible. Almonds, by the way, are close relatives of peaches but, through a genetic mutation, don't produce cyanide, so the seeds are edible.

The urban myth that swallowing an apple seed will kill you is based on fact, but blown out of proportion. An apple seed produces cyanide when digested, but since its seed coat is so tough, it passes through the digestive system intact (and undigested). Swallowing a few whole apple seeds is not a serious problem. Your body can even detoxify the small amount of cyanide released should a seed or two be damaged. If you were to crush a cupful of apple seeds (or peach seeds or cherry seeds) so that the protective seed coat is breached and then swallow them, you are likely to be fatally poisoned.

## Sweet Fleshy Fruits

Not all fleshy fruits are created equal. During the summer blueberry season, Appalachian slopes are covered by people with pails brimming with the season's bounty. In contrast, have you ever seen anyone pick up the attractive red berries of flowering dogwood and turn them into jelly? The main difference between these two types of fleshy fruits is their sugar and fat content, which makes them attractive (or repulsive) to different types of animals.

Fleshy fruits rich in sugar taste good to mammals. As mammals ourselves, we respond to these fruits, cultivate and enjoy them. Sugary fruits native to North America include blueberries, huckleberries, cranberries, raspberries, blackberries, red mulberries, paw-paw, persimmons, elderberries, some plums, and some cherries. The seeds of these fruits are either tiny and consumed with the fruit, such as those in blueberries, or large and usually avoided, such as those in plums. In fact, we are so enamored of sugary fruits that we have transported them around the world, and many now grow beyond their natural range. Apples, which are probably the most widely planted fruit tree in North America, are thought to be native to Kazakhstan, in south central Asia; oranges are native to southern China and northeastern India, peaches to western China and Tibet, and pears to Europe and the Near East.

These species have benefited by their association with us, for we have dramatically increased their range and number.

High-sugar fruits are among the first fruits to ripen in the summer months. They typically undergo a predictable series of color changes to indicate their ripeness. They are initially green and inconspicuous but turn red before they ripen and then (usually) turn blue or purple when fully ripe. These color changes alert animals with keen eyesight, color vision, and good memories to the coming bounty, keep them nearby, and thus assure the plants of a faithful means of dispersal. The easiest way to find ripe blackberries is to look for the bright red, unripe ones, and my young son easily remembers where all the best blackberry patches are.

Many of these plants are shrubs or small, sturdy trees. When loaded with ripe fruit, their limbs bend low to the ground. They are adapted to be dispersed primarily by mammals, such as bears, raccoons, squirrels and chipmunks, deer, opossums, and foxes. Although most of these animals can climb trees, they rarely climb very high. Birds also consume these fruits and disperse the seeds, especially those plants with small fruits and seeds that we usually refer to as berries.

While sugary fruits are low in protein and fat, mammals and birds eat a sufficient quantity of them to convert the surplus sugar into stored fat in preparation for the lean winter months. (Protein must be acquired elsewhere; omnivores get it from the animal portion of their diet.) While the plants are producing fruit as a way to disperse their seeds, they indirectly help their dispersers to prepare for the upcoming winter months of scarcity.

## Fatty Fruits? Spicebush and Dogwood

We rarely think of fruits as being high in fat and low in sugar, but such fruits are common. We just don't eat them. The high fat content makes the ripe fruit rancid, producing an unpleasant sour flavor to most mammals. Many birds, however, have poor senses of taste and smell, making them indifferent to sourness and sweetness. To migrating birds, high-fat fruits are especially attractive. Birds need a lot of energy from food to sustain their flight, but because they are flying, excess weight is undesirable. Thus, high-fat foods, which contain the most energy (there are nine calories per gram of fat and only four per gram of sugar or protein), are most desirable.

Spicebush, dogwood, sassafras, Virginia creeper, and several species of magnolia and gum produce high-fat, high-energy fruits that are particularly attractive to migrating birds. Spicebush fruits are 35 percent fat, flowering dogwood fruits are 24 percent fat, and the various species of *Magnolia* produce fruits ranging from 33 to 62 percent fat. Migratory birds, especially the thrushes, rely on these foods to sustain them during migration and, in turn, disperse the fruits over long distances. What better plant dispersal mechanism is there than a migrating bird? Indeed, most of these plants have a wide range of distribution up and down the Appalachian flyway.

Spicebush (*Lindera benzoin*) is widely distributed in eastern North America. It is a small shrub, rarely reaching ten feet in height, with stems an inch in diameter. It grows along stream banks in rich, moist forests, where it has easy access to water and some shade. It also prefers basic, rather than acidic, soils. In the Southern Appalachians, it is one of first of plants to show evidence of life in early spring, when its tiny yellow flowers emerge on bare, gray stems. The crushed leaves have a strong smell that reminds me of lemon-scented cleaning fluid, an interesting but not entirely pleasant sensation. The small, oblong red fruits are only about 1/4 inch long, but they stand out dramatically against the green leaves. In a New Jersey study, the fruit ripened during the first two weeks of September; by the end of October, 90 percent of the fruit had been dispersed.

Near Brevard, North Carolina, several high-fat fruits ripen in early fall. Spicebush ripens in early September, but the berries are very difficult to find; presumably they are eaten soon after ripening. Similarly, the fruits of sassafras appear in late August and disappear quickly. The blue fruits of sassafras are borne on red stalks that are prominent against the green leaves. The pink or red seeds of *Magnolia fraseri* ripen in early September and dangle from their cones, whereas the blue fruits of black gum (*Nyssa sylvatica*) ripen in late September; both last about two weeks. The red fruits of flowering dogwood ripen in early September and are usually gone by November. In addition to birds, I have observed gray squirrels feeding on flowering dogwood fruits.

Because high-fat fruits rot quickly, they must be consumed shortly after ripening, which corresponds with the fall migration season. To attract migrating birds that may be unfamiliar with the territory, the entire plant advertises its ripe fruits with a very early leaf-color change. The tree is said to be exhibiting a foliar fruit flag, which stands out conspicuously against the green backdrop. Foliar fruit flags, perhaps best exemplified by flowering dogwood, but occurring in several other species, including black gum, suggest to us that the fall foliage display is just beginning, but to migrating thrushes, they must look like what a fast-food sign looks like to a hungry driver!

Flowering dogwood (*Cornus florida*) is an attractive tree in both the spring and fall seasons. It is a tree of the forest understory, reaching about thirty feet in height. Although it grows and flowers in the shade of the forest canopy, it flowers more profusely with more sun, and is often used as an ornamental in suburban yards, as are Japanese flowering dogwoods. Perhaps best loved for its beautiful white spring display, its red fruits and purple-toned fall leaves are also striking. Flowering dogwood has been attacked by a fungal disease known as dogwood anthracnose (*Discula destructiva*). Because it is fungal, the disease grows best in shaded, damp areas. Thus, trees in moist forests are affected more severely than those in sunny locations.

## Other Fleshy Fruits: Sumac

Fruits that are high in fat are expensive for the parent plants to produce, and relatively few plants have become so dependent on migrating songbirds that they produce high-fat fruits. Some plants produce fall-ripening fleshy fruits that are low in fat (about 1 percent fat). Mountain ash, poison ivy, hollies, viburnums, junipers and cedars, sumacs, and greenbriers all produce these low-quality fruits, which are typically red but sometimes purple or blue. Several of these plants, such as the hollies (see chapter 5), viburnums, and mountain ash are used in the home landscape because they retain pretty fruits late into winter.

Since they are not high in fat, migratory birds ignore these fall fruits. Because they are not high in sugar, mammals do not gorge on them either. Thus slighted, they hang on the plant late into winter. (A study of mapleleaf arrowwood (*Viburnum acerifolium*) in New Jersey found that fruits ripened in late August, but by January, 70 percent still remained on the shrub and 20 percent had fallen off and remained on the ground.) Eventually, however, other food sources will be exhausted, and these lower quality fruits will be consumed and dispersed by hungry wintering birds and mammals.

Some of these lower-quality fruits, such as those of sumac, have a large indigestible seed covered by a thin layer of fruit, and the fruit typically remains on the plant well into winter. The entire sumac plant, however, shows an early foliar flag. Perhaps the sumacs present a foliar flag to tempt hungry but naïve migrants.

There are several species of sumac, also known as shoemake, in the Southern Appalachians, but staghorn sumac (*Rhus typhina*) and smooth sumac (*R. glabra*) are common. Staghorn sumac has hairy stems, while smooth sumac stems are smooth. Both grow along roadsides and other places where they receive a lot of sun. The compound leaves have seven to fifteen pairs of leaflets arranged opposite each other along the large midrib. The small, reddish

FIGURE 4-3

Smooth sumac (*Rhus glabra*) is a common roadside shrub. The reddish brown fruits are borne in dense clusters. The compound leaves have multiple leaflets.

brown fruits cover the branching flower stalk, which sticks up like the antlers of a deer (Figure 4-3). These edible fruits are high in vitamin C and were a staple for the Cherokee, who used them to treat colds and flu.

Poison sumac (*Toxicodendron vernix*) is not a true sumac but, rather, related to poison ivy. It is confused with sumacs because the leaves look similar: they are compound with seven to thirteen pairs of leaflets. However, poison sumac is easily identified by the flowers and white fruits that hang down in a loose cluster from the leaf axils of the plant. Poison sumac usually occurs in the swamps of the coastal plain. While reported in the Southern Appalachians, it is associated with rare bogs.

## Migrant and Resident Birds as Fruit Dispersers

The migratory thrushes, such as hermit thrush, veery, gray-cheeked thrush, Swainson's thrush, and wood thrush, are specialists on high-fat fruits. All move through the Appalachians during migration, some lingering longer than others. Resident thrushes, including American robins and eastern bluebirds, remain over winter in the Appalachians. Thrushes all have spotted breasts when young, and most are active hopping and walking along the ground. All eat fruits, especially during fall and winter, but they feed insects to their growing young in spring. During late winter, when other food is scarce, bluebirds outside my office window strip the horticultural hollies of their berries, finally dispersing the low-fat, low-sugar fruits.

Birds other than the thrushes are also important seed dispersers. Gray cat-birds (*Dumetella carolinensis*), which also migrate and eat fruits in addition to insects, are among them. They look like dark gray versions of mockingbirds, to whom they are closely related, and are common along the scrubby borders of forests and fields. They breed in most of the United States and winter in the tropics up into Florida. Their mewing call, usually issued when they are deep in tangled brush, gives them their common name.

Cedar waxwings (*Bombycilla cedrorum*) are residents that are important agents of short-distance fruit dispersal. They are fruit specialists that remain year-round in the Appalachians. Cedar waxwings are named for their habit of feeding on the berries of eastern red cedar (*Juniperus virginiana*), which are especially prominent during winter, and for the bright red waxlike drops on the tips of their wing feathers. With smooth yellow feathers on their belly fading into pretty gray on their back and dark highlights accentuating their eyes like those of ancient Egyptians, they are among the most beautiful of birds. They fly together in bouncy spurts from one tree to another, chirping brightly all the while. When a flock of these beauties descends on an American holly or mountain ash, it is bare of berries when they leave.

Even birds that are not specialists on fruits take advantage of fall's bounty. For example, pileated woodpeckers surprise me each fall when they dangle from thin vines to eat the berries of poison ivy, Virginia creeper, and fox grape.

## Dry Fruits: Oaks, Hickories, and Chestnuts

The eastern deciduous forests of the Appalachian Mountains are dominated by trees that produce dry fruits. Nuts are also fruits but are dry and woody instead of fleshy. The dried remains of the flower are sometimes visible on nuts such as chestnuts, and the caps, husks, or burs from which they fall are also part of the fruit.

As with yuccas and their pollinator moths (see chapter 3), nut trees have developed a mutualism with their seed dispersers. These trees include oaks, hickories, and walnuts, and the animals that disperse their seeds are primarily squirrels. The trees produce nuts in huge numbers, more than can possibly be consumed. They produce these windfalls irregularly, sometimes going several years without producing any nuts, a phenomenon known as masting (see chapter 1).

Oaks can be divided into two basic groups: the white oak group and the red oak group. The lobes of white oak leaves are rounded, but those of red oaks are sharp. The veins of red oak leaves also extend beyond each lobe to make a small bristle. Oaks in the group, including the white oak, and the chestnut oak, mature their acorns in a single year, dropping them in the fall. By contrast, red oaks, such as the common Appalachian species of northern red, scarlet, and black oaks, require two years to mature their acorns, so that a good spring one year produces heavy crops not in the fall of that year but one year later. Both groups produce tannins in their leaves and nuts, which protect the thin-shelled seeds from fungi and make them astringent.

Hickories, the other major group of nut-producing trees, are also diverse in terms of the number of species. Because their thicker-shelled seeds contain less tannin, we more often cultivate them for food. Pecans (*Carya illinoinensis*), for instance, are planted far beyond their natural range of the lower Mississippi valley, and many cultivars have been selectively bred for desirable qualities. Similarly, black walnut (*Juglans nigra*) has been bred for sweeter seeds enclosed in thinner shells than the wild type, and the uncommon white walnut or butternut (*J. cinerea*) of the Appalachians is highly desirable for its delicious nuts.

The American chestnut (*Castanea dentata*) used to be the dominant forest tree of the Appalachians, but it has been eliminated from the canopy by an introduced fungus, chestnut blight (*Cryphonectria parasitica*). It is estimated that 25–50 percent of all Appalachian trees were American chestnuts before the blight arrived in the northeastern United States in 1904; the blight first arrived in the Southern Appalachians in 1912 and spread rapidly during the 1920s. American chestnuts were effectively eliminated by 1940. Stump sprouts still emerge from the roots and may reach twenty feet in height, but they eventually succumb to the blight as well, rarely getting large enough to produce nuts. One of the few locations where American chestnuts regularly produce nuts is along the Blue Ridge Parkway near the Mount Pisgah Inn and the junction with Highway 276.

Chinese chestnuts (*Castanea mollissima*) are commonly found around old farm fields, where they were planted as a replacement for the magnificent forest trees, but they are smaller and thus more suited to domestic production. They are resistant to the blight but can harbor it; the blight was introduced to New York on a shipment of Chinese chestnuts. The chinquapin (*C. pumila*) is a native small tree or shrub that is less common than American chestnut stump sprouts. It has some immunity to the blight and frequently produces nuts before succumbing. It tends to grow in dry sites and produces small nuts. The underside of each leaf is furry to the touch.

## Squirrels as Seed Dispersers

Both eastern gray squirrels and eastern chipmunks are dependent on nut trees to sustain them, and the trees, in turn, rely on the animals to disperse their seeds. When nuts are plentiful, the animals cache the surplus. Chipmunks typically store their seeds and acorns in a single cache in their tunnel system. They spend the winter in hibernation underground, occasionally awakening to eat from their stored larder. Nuts that are not consumed, either because a chipmunk stored an excess or because it did not survive the winter season, may sprout from these caches. Gray squirrels, by contrast, cache in several locations outside their nest, shallowly burying individual acorns under the leaves. They are much more important agents of dispersal because they plant the seeds in many different locations. Again, those seeds that survive may sprout at some distance from their parent tree, and it is this dispersal that the tree depends on. Large acorns and nuts cannot be carried very far by the wind, and some trees, notably black walnuts, produce chemicals that inhibit their own seedlings from geminating in close proximity. (These chemicals are also inhibitory to many other plants, especially those of the nightshade family; tomatoes and potatoes will not grow near a walnut.)

Chipmunks (*Tamias striatus*) are ground squirrels, and while they often climb trees, they build their nests underground, where they hibernate for short periods in winter. An adult chipmunk builds its own burrow system and occupies it for the rest of its life. Chipmunks occur up and down the Appalachians, but not in the coastal plains from North Carolina to Alabama, and are most common in forested locations. Their small, brown bodies are about six inches long, and their tails are three inches long. They have thin white "racing stripes," with black stripes above and below them, on either side of their body. As they scurry about collecting acorns and leaves to line their winter nests, they hold their tail nearly upright.

Gray squirrels (*Sciurus carolinensis*) are ubiquitous. They occur from the mountains to the coastal plain, inhabiting both city parks and wilderness. Most of their body is gray above, and their belly is white. In Brevard, North Carolina, as well as in several other locations around the country, a white color morph exists. About 25 percent of the squirrels in the city's limits are white with a narrow gray stripe down the center of their head and back (see Plate 37). They are not albinos, for they have the gray stripe and dark eyes, but there is a genetic basis passed on to their young.

Gray squirrels are impressive acrobats, jumping from limb to limb across gaps that make me gasp. Sometimes, they miss their intended target and fall until they catch another branch farther down the tree. Because they are lightweight, and their spread-eagle legs and bushy tail act like a parachute, they usually survive a fall from a treetop as long as they land on forgiving ground. If they hit pavement, it can kill them. Gray squirrels are at ease moving up or down tree trunks, always head first. Their back legs can rotate the way our arms do, allowing their feet to point backward or forward to give them a good grip on the tree. Most other mammals, such as cats and raccoons, must back down a tree trunk because their hind limbs do not rotate.

Gray squirrels usually produce two litters of between two and five young, one in early spring and another midsummer. The second brood may stay with the mother through the following winter. They nest inside hollow trees or build dreys in treetops. Although they look like large bird nests, dreys are completely covered and have only a small side entrance. In addition, they are built with a high percentage of leaves, which most bird nests lack.

Squirrels handle the acorns of white oaks and red oaks differently. White oak acorns are low in tannins and germinate shortly after they fall from the tree. Squirrels tend to eat these acorns right away rather than cache them, but if they do cache them, they first nip the end of the acorns, which kills the oak embryo and prevents the nut from germinating. Squirrels are more often predators of white oaks rather than dispersers. By contrast, acorns of red oaks, which are high in tannins, do not germinate in the fall after dropping from the parent tree. Instead, they germinate in the spring. Squirrels cache proportionately more red oak acorns and rarely damage them because they normally last the winter. Tannins prevent the growth of fungi and bacteria, more important for the long-lived red oak acorns, but also make them taste bitter. (Native Americans, who ate acorns, also preferred the more palatable acorns of white oak.)

## Bird Migration: Kinglets and Nightjars

With the first cool weather of fall comes wave after wave of migrating birds coursing south along the Appalachians. The mountain chain is an important flyway connecting the entire East Coast of North America. To a bird hatched in Nova Scotia, the mountains are like an arrow directing it to the Gulf of Mexico. The bird then either flies straight across the Gulf or hops along the coast into Mexico and Central and South America. Ruby-throated hummingbirds (see chapter 2) and the migratory thrushes, for instance, fly nonstop over the Gulf of Mexico, whereas many warblers move slowly along the coast, sometimes straying farther west to fascinate bird-watchers there.

Birds that are primarily year-round insect feeders typically migrate to the tropics during the winter months because their food supply is greatly diminished by winter's cold. Because most insects overwinter as eggs or pupae, which are inactive and hard to find just at the time of year that small, warm-blooded birds need to find enough food to stay warm, small, purely insectivorous birds move farther south to survive. In moving south, however, they likely encounter more competition with other migrants as well as with resident birds. As soon as warmth and food returns to the north, the birds return as well. They move north to the temperate zone during summer because of the impressive increase in insect prey during the short season of abundance, which corresponds with their nesting season. They exploit the exploding insect populations to fill the bellies of their chicks.

Kinglets are among the hardiest of insectivores, moving merely from the Canadian boreal forests into the coniferous forests of the southern states. Golden-crowned kinglets (*Regulus satrapa*) are present in the Appalachians year-round but are more visible during the winter months, when pairs of these tiny, warblerlike birds chitter among low-growing shrubs searching for insect morsels. Their southernmost breeding location is in the spruce-fir forests of western North Carolina. Because they are bold and work low shrubs, it is easy to identify them by their olive-gray bodies, white wing bars, and white eyebrow, but the yellow crown patch that is exposed when they tip toward the observer truly gives them away. Males have a central orange stripe in the yellow crown. Ruby-Crowned kinglets (*R. calendula*) usually breed in Canada and winter in the southern United States, appearing in the Appalachians mainly during migration. They have white eye rings rather than the white eyebrow stripe, and males have a rarely visible red crown.

The insectivorous whippoorwill, chuck-will's-widow, and common nighthawk, which belong to the family of birds called nightjars, also migrate as their prey becomes less abundant. Nightjars feed on insects captured on the wing, a rare sight indeed in winter. Nighthawks, in particular, are long-distance migrants that migrate in huge flocks, soaring high above farm fields golden in the waning rays of afternoon sun, swooping and plunging as they execute astounding aerobatics. Described as "hawking" after insects, their awe-inspiring dives to earth and deep wing beats as they regain the firmament are all about catching insect food. Common throughout most of North America during the summer, they move to South America in winter.

The courtship of common nighthawks (*Chordeiles minor*), also called bullbats and thunderbirds, is a spectacular affair. The male flies upward until nearly out of sight, then, fast as an arrow, plummets toward earth in a nosedive. An instant before hitting the ground, the bird opens its wings and pulls out of the dive, and a deep resonant "boom" reverberates in the air. It is this boom as well as the birds' frequent appearance ahead of thunderstorms that accounts for the common name of thunderbirds. They are mistaken for large (bull) bats because they fly at dusk or at night, even hawking around street lamps in cities.

The nightjars are named for their vociferous calls that jar the night. Whippoorwills (*Caprimulgus vociferus*) are birds of the Appalachians and northeast, whereas chuck-will's-widows (*C. carolinensis*) are southeastern in distribution. Their eponymous calls advertise their territories to each other, and since they are active at night, their calls are heard at night. Hundreds of rhythmic calls are repeated in a row; the current record seems to be held by a whippoorwill that called its name 1,088 times, as recorded by a human insomniac. They don't migrate as far as nighthawks. Whippoorwills winter on the southeastern coast and chuck-will's-widows winter in Central America and the West Indies.

Up close these birds look as strange as their name would suggest. Nightjars have huge mouths and tiny beaks that seem to be an afterthought. Stiff feathers that look like mammalian hairs jut out around the edges of their mouths. The whole effect is that of a baby bird's wide gaping mouth surrounded by a sparse, wild moustache. Nightjars are also called goatsuckers, reflected in their family name, Caprimulgidae ("capri" is Latin for goat; "mulg," to milk). Legend has it that the birds drink the milk of goats, but they do not. The bird's surprisingly large mouth, which is reminiscent of a baby bird waiting to be fed or a suckling child, and its flight at dusk around fields and livestock, where it hawks for the insects that are plentiful there, are likely the sources of the legend.

For birds with such complete mastery of the air, nightjars have a very solid link to the earth. They nest directly on the ground in slight depressions they have scooped out in the leaves. The birds are the drab, uneven brown of last year's leaves and blend in so perfectly with the earth that they are nearly impossible to see. They rely on their camouflage to protect them during the day as they rest motionless on the ground. Their feet are tiny, so when they do sit in trees, they usually sit lengthwise along a large branch rather than crosswise on a small one, as most birds do.

## Hawks: Migrants and Residents

For the most part, migrating birds such as hummingbirds, thrushes, warblers, and nightjars are following their food source. By contrast, birds that are generalist feeders, especially those that eat seeds as well as insects, rarely migrate. Our common backyard feeder birds, such as chickadees, woodpeckers, sparrows, tufted titmice, and northern cardinals (see chapter 5), are able to find food throughout the year, and they do not migrate. Crows are known for their ability to find anything edible and are also permanent residents.

Hawks, however, are puzzling with regard to migration. A few eastern species of hawk are migratory, others are permanent residents, and some individuals of some species are somewhat migratory! For example, individual red-tailed hawks (*Buteo jamaicensis*), Cooper's hawks (*Accipiter cooperii*), and American kestrels (*Falco sparverius*) usually leave the northern parts of their range, in Canada, to winter farther south, but most southern individuals of the species are permanent residents, meaning that individual birds stay put on their breeding territories throughout the year. These permanent residents probably displace the northern migrants far to the south, beyond the range of occupied territories. On the other hand, sharp-shinned hawks (*A. striatus*), or sharpies, are mostly migratory, with most individuals moving from north to south during the fall, so that southern bird-watchers may see them all year long, but probably not the same individuals. Broad-winged hawks (*B. platypterus*) are fully migratory, all members of the species moving from the eastern United States well into South America to overwinter. Because individual hawks are hard to distinguish visually from one another, sophisticated methods of marking and tracking are needed to follow individuals. The difficulty and expense of such research, however, has inhibited our understanding of their migration patterns.

What can explain these different patterns of hawk migration? It is difficult to tie their migration to the pursuit of food, as with other migratory birds. Migratory broad-wings and resident red-tails consume primarily mammals, and migratory sharp-shins and resident Cooper's eat birds, and since birds and mammals are present year-round, they don't necessarily need to migrate (owls, for example, the nocturnal equivalent of hawks, do not migrate). Even the American kestrel, the smallest eastern hawk, is typically a permanent resident, although it favors insect prey that is less abundant in winter.

Hawk migration is likely related both to prey abundance and to competition between species. Although primarily mammal eaters, broad-wings prefer the larvae of large moths during late summer, and this prey source disappears in winter. Similarly, migratory warblers and thrushes are an important component of sharpies' summer diet but are absent in winter. Just when prey is generally less abundant (few insects, no migratory birds, hibernation of ground squirrels), more hawks of many species are concentrated in the

Perhaps the prime location for observing the fall migration of hawks is Hawk Mountain Sanctuary in Pennsylvania. The site was purchased by conservationists in the 1930s to protect the migrants from hunters, who killed hundreds of them for sport. Many high ridges along the Blue Ridge Parkway are also excellent locations, with the premier site at Rockfish Gap in Virginia, at the parkway's northern terminus. Hawks travel along ridge lines, where mountains drop off precipitously into flat country, such as at Caesar's Head State Park in South Carolina, because winds are deflected sharply upward, providing lift for big hawks. Of all the species of eastern hawks and eagles we observe in migration, kettles of broad-winged hawks are perhaps most impressive. Kettles form when hundreds of hawks swirl around a rising column of warm air, called a thermal. When the birds reach a certain altitude, they glide off downwind, often without ever flapping their wings. Most broad-winged hawks migrate around the middle of September, along with bald eagles, ospreys, and kestrels. Sharp-shinned and Cooper's hawks generally migrate in October, and red-tailed hawks typically migrate in November.

southern part of the United States. In this situation, broad-winged hawks probably compete with red-tailed hawks for mammalian food, and in a contest with the larger species, the broad-wings lose. Similarly, Cooper's hawks have an advantage over smaller sharpies. There are documented instances of both red-tailed hawks and Cooper's hawks killing and eating other hawks, and both species aggressively defend their permanent territories. Perhaps it is this intense competition in the southern wintering grounds that cause some hawks, both as species and as individuals, to avoid the southern range altogether and to head straight to South America. There is more to learn about hawk migration patterns.

Although unrelated to hawks, turkey vultures (*Cathartes aura*) have a similar, puzzling migration pattern. They are permanent residents in the South, but northern birds migrate as far as South America. Vultures, however, are scavengers, and it seems surprising that a scavenger would have a harder time finding food in winter, when starving mammals would seem to provide an endless food source. If, however, the dead animal is frozen solid, it would be difficult to consume.

## Resident Screech and Barred Owls

Owls are resident species, remaining within their territories year-round. Since they communicate with each other with their calls, owls are most vocal while establishing their territories. Eastern screech owls (*Megascops asio*) are usually heard from late August through October. Screech owls are named not for the sound they make, but because of the sound some people make—a screech—when hearing them! Like the whinny of a horse possessed by demons, the screech owl's call modulates from a soft cry to a quavering sob. A panicked Brevard College student once claimed in a campuswide email that a ghost followed him as he ran to his dormitory one September night, but I reassured the student body that he heard an owl instead.

Screech owls occur throughout the eastern two-thirds of the country. In the northern part of their range and throughout the Appalachians, they are most often reddish in color, but southern owls are usually gray. They are not as shy as many of the other owls, and they even occur in urban parks. They can be found in heavily forested areas as well as open farm fields.

Screech owls feed on mice and large insects. In turn, they may become food for the larger owls. Once a great horned owl or a barred owl begins to hoot, the diminutive screech owl gets quiet! As cavity-nesters, screech owls normally nest in hollow trees. They will, however, readily use constructed boxes. You may be able to attract a pair to your property if you build a nest box. On early fall evenings, you'll be treated to the strange song of this small owl. Like me, you might smile with the knowledge that the population of mice living in your woodshed or garage will soon decrease while your owls raise their young.

In autumn, the only other owl you might hear is the barred owl (*Strix varia*). Its call is loud and emphatic and sounds alot like "Who cooks for you? Who cooks for y'all?" A long, drawled "y'all" at the end of the call is characteristic of the female; the male's call ends more abruptly. Barred owls are able to make an incredible range of other sounds, many of which disturb the peacefulness of an evening's walk or camping trip with their otherworldly quality. One of these calls sounds like the high-pitched yapping of a small dog, but one located above, in a tree.

Barred owls prefer deep forests and occur in coastal swamps as well as high mountains all along the eastern half of North America. Besides uncommon barn owls, they are the only Appalachian owls that have brown eyes rather than yellow ones. They are named for the bars that run across their chest, contrasting with the streaks that run vertically on their belly. Great horned owls (*Bubo virginianus*), the other "hoot owls," call in winter and early spring (see chapter 5).

## Bats

Too many people think of bats as filthy, rabid bloodsuckers whose sole desire is to fly into and tangle your hair. But how far from the truth! We misapprehend them because their natural history is so different from our own. Bats are nocturnal, which may be the reason for their bad reputation, because nocturnal animals differ from diurnal animals in basic biology. Bats use their sense of hearing rather than sight to catch food, navigate, and locate other bats. In fact, blind bats can operate normally, but deaf bats are soon dead bats, for they can neither feed nor fly.

Echolocation is bats' primary sense. As they fly, bats open their mouths and emit high-pitched shrieks that are beyond the range of our hearing. Their strangely elongated noses help focus the outgoing beams of sound, in the same way that a megaphone can focus a human voice and project it over a longer range. Their large ears receive the returning echoes from objects in their paths, concentrating the incoming sound waves as a cupped hand or ear trumpet will do. Bat brains use those echoes of sound to generate a picture of the world in the same way that our brains use signals from the retinas of our eyes. The images created by echolocation allow bats to "see" in the dark. They can avoid obstacles as thin as a single human hair, not to mention an entire head of hair!

Imagine navigating your environment with sound instead of sight. Some exceptional blind people can echolocate by listening to the echoes of a tapping cane or clicks of their tongues. Using echolocation, they can locate and identify objects.

Bats are more active at night because their insect prey, mainly moths, beetles, and flies, is also more active at night. Some tropical bats eat fruit, blood, fish, and other animals, but all Appalachian species are nocturnal insectivores. By flying at night, bats avoid competition with most birds for those tasty and nutritious insect morsels.

Bats are significant controls on insect populations because they must eat enormous quantities of food to power their high metabolism. Their normal temperature is about 100° F and, when they are flying, their heart rate is 1,000 beats per minute. Most bats eat half their body weight in insects every night. For instance, a colony of endangered gray bats (*Myotis grisescens*) near Chattanooga, Tennessee, is estimated to eat 220 tons of insects each summer from the surrounding area.

Because their insect prey is greatly reduced during winter, bats, like insectivorous birds, must either migrate to warmer locations abundant in insect food or hibernate and slow their metabolism to a level that can be sustained by their fat reserves. Most Appalachian species hibernate in caves. During hibernation, their temperature drops to that of the hibernation site, usually around 50° F. Their heart rate drops to fewer than fifteen beats a minute. Even with such a slowed metabolism, bats lose about one-quarter to one-half of their weight during hibernation, which lasts five or six months.

While it might seem that bats, because they have wings, should migrate rather than hibernate, few species do. Bat flight is fluttering and feeble compared to the strong, direct flight of birds. Echolocation would probably not serve them well in migration, for it works best over short distances. Most migrating birds fly high and use the stars, the setting sun, and large landmarks to direct them, none of which can be detected with echolocation.

Bats are mammals. They nurse their young with mammary glands and are fur-covered. They fly using their hands. Their wings are basically very long fingers with skin stretched between them. A skin-flap extends from the little finger down to the ankle, and another flap of skin joins the two feet, which increases the surface area of the flight membrane (Figure 4-4).

Contrary to popular belief, bats are quite hygienic. They groom themselves like cats and are fastidious about cleaning their wings. Although it is correct that their roosting caves are often layered with their droppings, the guano builds up on the floor where we scrabble, not on the ceilings where they live. The ceilings are clean, providing secure toeholds for adult and young bats. Most hibernation caves have probably been used by bat colonies for thousands of years, for even bats that have been moved over two hundred miles away return to their home caves. Humans have harvested the concentrated guano as a source of nitrogen for fertilizer and gunpowder. It is these human collectors of guano who become filthy, not the bats!

The two most common bats in the Southern Appalachians are the little brown bat (*Myotis lucifugus*) and the eastern pipistrelle (*Pipistrellus subflavus*). During an hour-long event one July in western North Carolina's Pisgah National Forest, where bats were captured, measured, and released, we identified eleven little browns and two pipistrelles. Both species hibernate in caves. While little brown bats commonly form maternity colonies made up of hundreds of mothers and young inside buildings, pipistrelles rarely enter dwellings. The maternity colonies, by the way, will leave the building at the end of summer.

Bats have a low rate of reproduction, usually giving birth to only one off-

FIGURE 4-4

Bats that use echolocation have large ears but tiny eyes. They can hang by their back feet or their free thumbnails. This little brown bat (*Myotis lucifugus*) was rescued from a smoky chimney. It had crawled into a water bowl before this photograph was taken; some of the darker areas of fur are wet.

spring per year. The young are born in May or June and can fly in two to five weeks after birth, depending on the species. Most mature in the year after their birth, mating in autumn. They are long-lived, especially for their size, with a life span of thirty years. In comparison, similar sized mice are reproductive within thirty-five days, can produce several litters every year, each with up to seven young, and typically live for only one year in the wild (mice in captivity live as many as eight years). Bats, on the other hand, are active for only about half the year, so perhaps they should be credited with only a fifteen-year life span!

In the Appalachians, bats rarely carry rabies but should never be picked up with bare hands. Like most frightened animals, they will bite if restrained, and any bat on the ground during daylight hours is not acting normally and may be ill. Occasionally a bat becomes trapped in a house when it's searching for a roosting site. If one finds its way into your house, don't panic! Simply open the windows and doors and let it find its way out.

Perhaps the most infamous bat is the legendary vampire bat (*Desmodus rotundus*) of Central and South America and Mexico. It does, indeed, drink blood from its victims, which are usually cattle or goats but also sometimes

Bats have much more to fear from humans than humans do from bats. Many species of bats are federally endangered or of special concern because their numbers are declining throughout their range. They have few natural predators, and human activities are to blame for the decline of most species.

Disturbing maternity roosts and hibernation sites in caves or abandoned buildings is one way we endanger bats. When maternity roosts are disturbed, the young bats can fall to their deaths as their disoriented mothers flutter and fly. Each dropped offspring is the entire reproductive effort of a female for the year. Disturbing bats during hibernation causes their metabolism to increase, and a single significant wakening is equivalent to about three weeks of hibernation in terms of their fat consumption. They have little excess to spare and may starve before their food supply resumes.

Another factor in their decline may be pesticides. Bats consume enormous quantities of insects, many of which likely carry pesticide residue in their bodies. Over the thirty-year life span of the bat, pesticides may build up to levels that are detrimental to their reproduction.

Destroying a maternity colony kills adult breeding females as well all the young produced that year. Don't do it! If bats have taken up residence in your attic, it must have a direct opening to the outside. Let the bats show you the hole when they fly out at dusk (or in at dawn), and then cover it *after* the bats leave in the wintertime, and thank them for locating the cause of your high heating bills.

humans. The bat lands on the ground and drags its body and crumpled wings clumsily toward the sleeping victim. (Bats have sacrificed walking in favor of flight; imagine trying to walk with both legs tied together and both arms tied to both legs!) The bat nips the sleeping animal on its ankle or toe, and then laps up the oozing blood with its tongue. Its teeth make a swift, sharp slice, and its saliva contains an anesthetic, so the victim feels no pain and rarely awakens. The vampire's saliva also contains an anticoagulant, sometimes called "draculin." Two other species of vampire bats also occur in Central and South America, but they feed mainly on birds. None of these vampires occur in the Appalachians, not even in Transylvania County, North Carolina!

## Bats and Gargoyles

While visiting a local castle, the magnificent Biltmore House in Asheville, North Carolina, I had a close look at some of the gargoyles adorning the cornices. They look like bats! Gargoyles are almost always winged, with long ears, contorted faces, and fangs. Bats have prominent ears because their hearing is their most important sense. The weird faces, with fleshy outgrowths around the mouth and nose, help to focus echolocation sounds. Bat eyes are tiny and their faces seemed pinched because they don't need the additional input afforded by good eyes placed far apart in an open face. The bats' spiky teeth are used to catch the insects they eat, and their mouths are usually open while they hunt. Conjure up a scary beast to put a little thrill into your Halloween this autumn: give it wings to fly silently through the night, elongated ears the better to hear you with, a wrinkled face with beady eyes and prominent nose, and stiletto teeth bared into a grimace. Now you have a bat, and all those scary features are simply the way it catches mosquitoes and moths by using its hearing. Our imaginations have run amok with one of nature's many forms.

## Migrating Monarchs and the Foods that Support Them

Insects, like bats, rarely migrate even though both have wings. Most adapt to changing seasonal conditions with their rapid and complex life cycles. Their wings, however, provide an opportunity for long-distance flight, and at least one insect has adopted migration as part of its natural history. Monarch butterflies (*Danaus plexippus*) have become emblematic of long-distance migrants and the challenges they face. They travel thousands of miles over which they must find food, fend off predators, and navigate environmental obstacles.

Monarchs in the eastern part of North America winter in the mountains of Mexico. There the perfect combination of moist and cool conditions allows the monarchs to slow their metabolism and survive for over six months of the year. Huge groups of hundreds or even thousands of butterflies huddle together, covering the trees in a profusion of flickering orange and black forms. It is astonishing that a creature so small and so fragile should be able to fly thousands of miles, but millions do. A monarch tagged near Toronto, Canada, was found, some weeks later, at the winter roost in Mexico.

Not every monarch migrates, however. Each summer, several generations of monarchs hatch from eggs, grow into caterpillars, transform into chrysalises (Figure 4-5), and metamorphose into butterflies. These adults mate and lay eggs, and the next generation begins. Caterpillars that hatch in the waning days of summer, however, are physiologically different from the other generations. As soon as they metamorphose into butterflies, they begin the

FIGURE 4-5

The larvae of monarch butterflies (*Danaus plexippus*) are boldly patterned because they are poisonous. Just before the larva pupates, it hangs upside down from a leaf with its head curled forward into the shape of the letter J. The skin splits and then falls away, revealing the beautiful green chrysalis with its row of black and golden spots.

migration south. They do not mate and lay eggs but instead feed and store fat. Once they reach their winter destination, they eat rarely, if at all. After more than six months in their winter roost, these monarchs return north, but they fly only far enough to find milkweeds, where they lay their eggs and then die. Several generations are required to reach the northern limits of their range before the final, migrating generation is born and the cycle is renewed.

To avoid predators, monarch caterpillars feed on milkweed plants, whose poisonous compounds make the butterflies distasteful. One autumn I observed a broad-winged hawk that was sailing south with the butterflies grab a monarch from the air with its foot, shove the butterfly into its mouth, and promptly spit it out again. Remarkably, the butterfly appeared to be unhurt and resumed flying. Monarchs advertise their distastefulness with bold orange colors and distinctive black stripes, and other butterflies mimic this color pattern for protection (see chapter 3).

Milkweed plants (*Asclepias* spp.) produce milky latex that oozes from the stems and midribs of the leaves when they are damaged. The latex of milk-weeds discourages grazing by most animals. Not only is it sticky and volumi-nous, but it is full of poisons. The poisonous compounds are toxic resinoids (galitoxin) and cardiac glycosides that if ingested by livestock, will cause weakness, seizures, difficulty breathing, and death.

Only a few specialized insects like the monarch caterpillar have adapted to eat milkweeds. They minimize the latex flow by clipping the midrib of a leaf close to its connection to the stem. Then they can chew on the edges of the leaf farthest from the stem while the latex drips away. In addition, their bodies have adapted to the chemical compounds. Insects that eat milk-weed may have to expend energy to prevent being poisoned, but they don't have much competition for their food and it makes them toxic to predators. Milkweed bugs (*Lygaeus kalmii, Oncopeltus fasciatus*) are bright red and are as toxic as monarchs. The caterpillars of the milkweed tiger moth (*Euchaetias egle*), like those of the monarch, are a distinctive orange, black, and white, and the adult moths advertise acoustically to their bat predators (see chapter 3). While these insects can occur on more than one species of milkweed, each has a preference, which is probably related both to the level of poisonous compounds in the plant and to the digestibility of the foliage. For example, poke milkweed (*A. exaltata*) commonly hosts the beautiful caterpillars of the milkweed tiger moth, but I've never observed monarch caterpillars feeding on it. Common milkweed (*A. syriaca*) is probably best known as a food source for monarchs, but swamp milkweed (*A. incarnata*), also an Appalachian spe-cies, is reportedly more favored.

Milkweed flowers are not toxic, and their nectar is relished by many butter-flies and other insects. Each flower has a characteristic shape that is strange enough to warrant a close examination. The lower portion of each of the five petals bends down sharply, while the upper part fuses with the filaments of the anthers to form a hood. The hoods surround the single, central stigma. Each hood contains attractive nectar as well as pollen packed into pollinia like those of orchids. Insects that visit the flowers for their nectar may pick up a pollinium on one of their legs and transfer it to the stigma of another flower. Successful pollinators are usually large butterflies or bumblebees that are strong enough to dislodge the pollinium and fly to another flower with it; smaller insects may actually be trapped by it.

Monarchs are one of the few species of insects in which the sexes can be determined at a glance. The black lines on the hind wings of male monarchs are thinner than those of females, and they thicken to form a small black spot in the center of each hind wing (Plate 45).

One of the best places to observe monarch migration is the Cherry Cove overlook on the Blue Ridge Parkway, near mile marker 416. There the monarchs pass through a natural gap in the mountains that funnels many of those flying from more northern states into a narrower stream. Anytime after the air warms up, from midday to late afternoon, is a good time to see them, and the largest number of butterflies is usually seen during the last two weeks of September. Many areas in the Southern Appalachians, especially where unmown fields are allowed to flower, provide resting and feeding stops for butterflies to congregate en route to points south. To stand on a ridge amidst a monarch migration, as butterflies suddenly appear over trees to the north, sail by within arm's reach, then drop down over the ridge to the south, is to be swept up yourself in life's current.

During migration, monarchs use environmental conditions to their advantage, waiting on an early cold front to push down from the north to carry them south. When the weather turns cool or rainy, or the winds shift, they wait until conditions improve, which can result in some spectacular "fallouts." One unseasonably cool September evening, I glanced upward as I stood near the edge of a flowery field, and there among the tired, yellowing leaves of a grand old tulip tree was a cluster of orange-winged monarchs. Huddled together, they moved in the breeze as if the mass of them were a single, living organism; they were a kaleidoscope of color shifting against the azure afternoon sky and luminous leaves of the tree.

The fall migration of monarchs coincides with the fall blooming period of many composite flowers in the family Asteraceae. In September, shaggy fields burst into a blaze of color from the blossoms of yellow goldenrods, blue asters, purple ironweeds, and pink joe-pye weeds, and these sweet sources of nectar are often covered with feeding butterflies.

Composite flowers are among the most common flowers of fall. Each large "flower" is really a composite of many smaller flowers, hence the name, and a sunflower is a good example. Located in the center of the sunflower head are the fertile flowers, with stamens and pistils, and they alone produce the seeds in the central disk. Around the rim of the flower head are bright and colorful, but sterile, ray flowers. Their sole function is to attract pollinators. The union of many small flowers into one large, conspicuous, composite flower attracts pollinators and produces a large number of seeds.

Fall-blooming asters not only bear composite flowers, but many are large

While many people are attracted by the beauty of monarchs and intrigued by their migration, not all realize that they are also contributing to monarch decline. Both monarchs' winter roosts and summer food sources are affected by human activities. The highly localized winter roost in Mexico is threatened by timbering operations. Although the actual trees in which they roost are now mostly protected, the forest around them is disappearing, and the intact larger forest is what produces the climatic conditions the monarchs need to survive. As an example, winter storms in January of 2002 and again in January of 2004 killed large numbers of monarchs, and the severity of the storms' effects increased with reduced forest. Scientists are concerned that the butterflies may be unable to recover from these severe and increasing losses. Agricultural pesticides are also killing monarch caterpillars. Studies show that pollen from corn that has been genetically engineered to kill pest caterpillars also kills monarch caterpillars. The toxic corn pollen is easily distributed by wind onto the leaves of milkweed plants nearby. Furthermore, most farmers are also removing milkweed from their fields because it is toxic to their livestock, thus reducing the availability of food for monarch caterpillars.

plants, some reaching ten feet in height. Perhaps they are tall, with large flower heads, to lure migrating monarchs to pollinate them, just as some trees use early fall color to attract migrating thrushes to disperse seeds. Large plants are also able to capture and store more light energy than a small plant. Much of that additional energy can then be converted into the numerous seeds they produce. Most of the seeds from these tall plants are dispersed by the wind, just as some tall canopy trees depend on wind to disperse their seeds. The interrelationship between plant body size, pollinators, and seed dispersal is a fertile area for study.

## Ladybugs: Friends or Foes?

While monarchs migrate, ladybugs hibernate. There are several native species of ladybugs in the Appalachians, but an introduced species, the Asian multicolored ladybug (*Harmonia axyridis*), has begun to besiege homeowners on the East Coast and in the Midwest. As summer's warmth ebbs away, these ladybugs assemble in warm places to bivouac for winter. They are especially attracted to locations that are exposed to the winter's sun. In nature, they look for crevices in the bark of trees, but they are perfectly happy to find a narrow channel around a window that leads into a house, especially if the opening is on the sunny southern or western exposure. If one finds a good place to take refuge, it sends out the call and hundreds follow. They use pheromones, chemicals to which they are extremely sensitive, to locate each other.

Ladybugs are actually beetles, with large hard covers, called elytra, to protect their delicate flying wings. In Asian multicolored ladybugs, the elytra range in color from orange to pale yellow, and they may or may not have black spots. They are easiest to identify by carefully observing their white "neck," called the pronotum, which is just behind the tiny black head and in front of the colorful elytra. On the Asian ladybugs, the white pronotum has a black letter M (see chapter 5 and Plate 49). Our native species of ladybug also have black and white pronota, but none has the distinctive M.

The Asian ladybugs were purposefully imported for use in organic gardening because they and their larvae are effective predators of aphids; many garden catalogs sell ladybugs as biological controls. Like many other exotics, they have rapidly multiplied because our environment is similar to their native Asian environment, with the exception that here they have no natural enemies. Asian multicolored ladybugs are not causing any harmful environmental effects, but they have only been here about twenty years (1988 was the first documented specimen collection). Most folks with hordes of ladybugs inside their homes would argue that they are already pests.

Because they are predators of aphids and because they are native to Asia, there may actually be a surprising benefit to their introduction. They are one of the few predators of introduced hemlock woolly adelgids (see chapter 5); they even tear into the cottony egg masses to feed on them. Unfortunately, these ladybugs are only attracted to sunny locations, and since the majority of hemlocks grow along shaded streams, they have limited benefit.

If ladybugs have found a way into your house, they will be with you until spring, and if they plagued you last year, they will be back. If their tactics overwhelm you and a rampant horde invades, try vacuuming them up and releasing them outside. You can even take off the vacuum bag and leave that outside — the beetles will emerge from it in the spring. If you pick them up or sweep them, you'll notice that they bleed as a defensive mechanism. Be warned that the blood smells (and tastes!) unpleasant and will stain fabric. In addition, they also bite, but not very strongly. The only way to prevent them from overrunning your house is to caulk the cracks around windows and doors. Caulking will decrease your energy bills and is one way you can contribute to energy efficiency — a big plus for yourself and the environment.

So the ladybugs are really just pointing out something you need to address anyway.

## The Rise and Fall of Insect Populations: Woolly Bear Caterpillars

An insect is best understood as a complete life cycle—egg, larva, pupa, adult—and not only as the adult form. For most insects, preparation for winter usually involves the decline and death of the short-lived adult stage, whose main function is to leave behind eggs for the next generation. Most insects overwinter as eggs, and a few do so as larvae or pupae. Woolly bear caterpillars (*Pyrrharctia isabella*), for example, are so active and visible during the fall season because they are searching for a safe place to endure winter.

Woolly bear caterpillars are about two inches long and densely covered with stiff hairs, which make them look as woolly as a bear (Plate 46). They have a black head and tail and are brown in the middle. When the caterpillar is threatened, it curls into a tight ball with the bristly hairs sticking out in all directions to discourage predators, mainly rodents such as mice, from eating it. The woolly bear's body is divided into thirteen segments. According to legend, a long brown middle predicts a mild winter and a short brown middle predicts a harsh one. There are several methods for measuring the length of the brown section, but *The Old Farmer's Almanac* reports that most soothsayers count the number of brown segments.

The caterpillars eat a variety of plants, including grasses, clovers, plantains, and dandelions. These plants occur along roadsides and in open fields, which is why the caterpillars are seen so often. They seem to race across the roads and are, in fact, pretty fast. They have been clocked at four feet per minute, or about 0.05 miles per hour. By way of comparison, humans walk at approximately 3 miles per hour, but if we were scaled down to caterpillar size, our walking speeds would be similar. Why they are so conspicuously mobile, putting themselves at risk for predation, is a mystery. Their food sources are readily available, as are their overwintering sites, which are protected locations under leaf litter.

The woolly bear caterpillar is the larval stage of a tiger moth. There are several other tiger moths in the Appalachians, and many of their caterpillars are similar to woolly bears. Two common species are about the same size, are covered in stiff hairs, and curl up into a ball when threatened. The caterpillar of the giant leopard moth (*Hypercompe scribonia*) is black with reddish bands between its bristles that are visible when the caterpillar rolls into a ball. The yellow bear (*Spilosoma virginica*), while usually pale yellow, can range in color from white to dark brown.

In Banner Elk, North Carolina, where the best-known woolly worm festival takes place each year, the caterpillars are raced up a string, and the winner is used for predicting the severity of the upcoming winter. A special formula is used to "read" the winning worm. Its thirteen body segments are believed to correspond to the thirteen weeks of winter, and the severity of cold during each week is predicted by the color of each segment. The festival is usually held during the middle weekend in October. For more information about the festival, go to <http://banner-elk.com>.

Unfortunately, caterpillars can't really predict the weather. Rather, the caterpillars are actually reporting *previous* weather conditions because the amount of black on their body is determined partly by the temperature and level of moisture in their habitat during their early life. Woolly bears that are active in the fall were hatched from eggs laid during the summer. So while woolly bears' appearance might be affected by the weather, it is probably the weather of the past summer rather than the upcoming winter. In all fairness, however, weather patterns seen in the summer may be somewhat predictive of the following winter weather, as best exemplified by the El Niño phenomenon. The amount of black on a caterpillar also varies with its age (more brown appears with each molt), and not all caterpillars found on a given day are the same age. For example, immature woolly bears collected in the fall of 2005 in Transylvania County, North Carolina, had black head and tail sections that were about the same size as the brown midsection, while the full-grown ones had tiny black sections and a huge brown midsection.

The adult moth of the woolly bear caterpillar is about two inches long and yellowish in color. It has three rows of six black dots on its body. The moth must be as common as the caterpillar, but it tends to be overlooked because it is one of many pale, night-flying moths.

## Woolly Alder Aphids

November is truly the month of transition to winter, as most plant and animal activity comes to a halt. As the last of the leaves part company with the branches that produced them, other aspects of the forest are revealed. Forest denizens once obscured by the leafy green blanket now lie exposed to observant eyes and curious minds. One such concealed animal is the aphid. In particular, millions of woolly alder aphids (*Paraprociphilus tessellatus*) appear as a white, cottony covering on the bare branches of alder shrubs.

Common, or hazel, alder (*Alnus serrulata*) is a shrub that grows along stream banks, pond margins, and other wet areas. It has roundish leaves, mostly gone by November, and small cones that may persist through winter. Its leaves are somewhat similar to those of hazelnut (and witch hazel), hence one of the shrub's common names. The shrubs reach about fifteen feet in height and form dense, brushy thickets.

Woolly alder aphids make a white, waxy fuzz to cover and protect their soft bodies. Look closely at the white patches and you will see the many gray, soft, fat bodies of the insects underneath. Although the fuzz makes them more visible, it also makes them distasteful. Instead of getting a mouthful of nice, juicy insect, a predator first gets a mouthful of waxy, cottony fibers.

Many types of aphids are protected by ant guardians, and woolly alder aphids are no exception (Figure 4-6). These ants protect the aphids from predators and competitors. The ants rush out to bite and otherwise discourage anything that touches the aphids' alder, whether it is a human finger poking the branches or a caterpillar looking for a leaf to munch on. In exchange for their protection, the ants milk the aphids for honeydew, which is a sweet concentration of plant sap that the aphids excrete. Aphids suck up volumes of plant sap, extruding the excess sugar as honeydew while concentrating the nitrogen, which is needed to mature their offspring. When the ant guardian strokes an aphid with its antennae, the aphid, like the cow being milked by the farmer, releases a drop of honeydew to the ant as a food reward. Sometimes, however, the ants are unable to keep up with the aphids, and the honeydew drops onto leaves and twigs below the aphid colony. A sooty mold (*Scorias spongiosa*) is able to grow in the sweet honeydew and forms black, spongy masses.

Scattered among these woolly aphids are camouflaged predators. Truly wolves in sheep's clothing, the larvae of lacewings (*Chrysopa* spp.) prey on the aphids and cover themselves with the wool of their victims. It has been shown experimentally that if the wool is removed, the lacewings are more vulnerable to attack by the guardian ants. Larvae of some hover flies (family Syrphidae) and the caterpillar of the harvester butterfly (*Feniseca tarquinius*) are also predators on the aphids.

Aphids feed on plant sap by piercing the stem of the plant and sucking out the juice. Once in place, they remain on the host until fully grown, although they can detach and move around to more favorable areas. Like ticks on a dog, they parasitize the plant for its sap. Rarely do they move beyond the tree on which their mother placed them, unless they are members of the

FIGURE 4-6

Woolly alder aphids (*Paraprociphilus tessellatus*) are guarded by ants because they produce honeydew on which the ants feed.

last generation, born just before winter sets in. Only these aphids have wings and can disperse to other plants.

## Spiders and Their Insect Prey

Spiders are predators, and their primary prey is insects. As the number of insect prey grows through the summer months, so does the population of the predatory spiders. When dew is squeezed from the previous day's warm, moist air by the cool nights of fall, it settles on spiderwebs, rendering them visible. By the time fall arrives, the number and diversity of spiders, suggested by visible webs, are staggering. In one study in Great Britain, a hectare (2.4 acres) of grassy, undisturbed meadow supported 5 million spiders.

All those webs are positioned to capture insect prey. Since there are many different kinds of insects, spider webs are also diverse. Some spiders, such as sheet-web spiders, build webs that are mostly horizontal so that they trap insects as they fly or hop up from the ground. Orb-weaver spiders construct those magnificent vertical spirals that are most commonly associated with spiders. Quite a few spiders do not use webs to catch prey at all but hide in flowers to ambush insect pollinators or stalk through leaf litter and pounce on their prey like miniature tigers. If you spend an afternoon in a meadow, you'll see many of these different species, but to appreciate them, your attention must shift toward smaller forms of life.

Spiders use silk for more than just webs to catch prey. The original use of silk was probably for reproduction. All spiders encase their eggs in a silken cocoon. Not only does the silk protect the egg case from rainfall and from desiccation, but it also ameliorates other environmental challenges. The egg case can be glued in place to overwinter in a secure location. The silken cocoon keeps out small predators, and larger predators may find the silk distasteful. Male spiders also use a web as part of their reproductive cycle. They spin a small sperm web, deposit sperm onto it, and then take up sperm from the web into their pedipalps, which are clawlike structures located near their head. They use the pedipalps to transfer sperm to the female. Male spiders can usually be identified because their pedipalps are larger than those of the female.

Silk is also used by most spiders as a dragline. As they walk, they lay down a thread of silk, anchoring it occasionally to objects as they move along. When they jump or fall, the dragline catches them and they can control their descent by letting out more silk. They ascend by climbing the dragline. Similarly, in a dispersal process called "ballooning," young spiderlings climb to the highest nearby point and cast a dragline into the air. Breezes will pick up the tiny spiders, which ride the silk threads to a new location. Ballooning spiderlings can travel great distances over open bodies of water, colonizing islands or surprising boaters when the silken strands catch on rigging and glitter like golden streamers. Spiders also build underground nests of silk as daytime retreats or protected refuges in which to overwinter.

The spectacular orb webs are the most advanced natural use of silk. Different orb-weavers can often be identified by the particular shape and construction of their web, but the basic structure is the same. Each web consists of a catching spiral, radiating spokes that support it, a frame that anchors the web, and a central hub. Only the catching spiral is sticky. Other types of nonadhesive silk are used to build the web's frame, its spokes, and the central hub where the spider waits; spiders are able to produce many different kinds of silk that are used for different tasks. A spider avoids catching itself in its own web not by any magical antistick device but by its careful behavior. It simply avoids walking on the sticky, catching spiral. Instead, it moves around in its web by walking only on the nonsticky frame between the catching spiral and the central hub, a zone free of spiraling silk that increases the safe, nonsticky area and allows the spider to move between the two faces of the web. The large, thickened central zigzag that decorates the webs of some species is probably placed there as a warning to flying birds and running mammals to avoid the web. These big animals would destroy the web if they hit it, forcing the spider to spend time and energy producing silk and rebuilding the web. This is a particular hardship given that the spider has already been fasting while it constructs its web.

When prey hits and sticks to the web, spiders respond to its vibrations. Spiders are able to determine the size of prey from the vibrations alone, and they behave differently according to prey size. Large prey is bitten to kill it and then wrapped in silk, but small prey is wrapped and then bitten.

Although spiders are probably the best-known silk makers, a few of their arachnid relatives also make silk, and quite a few insects do. As arachnids, spiders are more closely related to ticks, scorpions, and horseshoe crabs than they are to insects such as bees, butterflies, beetles, and flies. Spiders make silk in their abdomens and release it from knobs at its tip, called spinnerets. Insects make silk in the salivary glands of their heads and release it from their mouths. Insects also use silk to enclose pupae (silkworms), construct nests (webworms), deploy draglines (inchworms), mark trails (tent caterpillars), and capture food (caddisfly nymphs).

Spider silk is a protein that is produced as a liquid but hardens into a solid as it is physically drawn out into a strand. It is as strong as nylon but more elastic. These properties of spider silk have drawn the attention of engineers who hope to manufacture a facsimile of it, but as yet they have been unable to come up with a product of similar quality.

All spiders produce venom to immobilize their prey quickly and begin the process of digestion. They use fangs to inject the venom, which begins to break down the body tissues of the prey so it can be sucked up by the spider. Eventually the spider discards the dry husk of the insect. Two spiders are dangerous to humans because we react strongly to their venom. The female black widow spider (*Latrodectus* spp.) has a smooth, glossy black body with a red hourglass on her abdomen. These spiders are common in protected locations such as piles of rock or wood, but they rarely come indoors. Their cobwebs are low to the ground to capture crawling insect prey. Their venom is a neurotoxin and affects both nerves and muscles; an antivenin is available. Rarely fatal, but usually agonizing, their bites can produce severe and painful reactions, including muscle spasms, abdominal cramps, and respiratory paralysis. The brown recluse spider (*Loxosceles reclusa*) does not occur in the Appalachians except in the extreme southern foothills of Alabama and

Georgia. Normally, they occur in the Midwest. They are pale brown with a dark marking on their thorax that resembles a violin. Since other spiders have similar markings, it is the brown recluse's six eyes rather than eight that positively identifies it. Their venom is hemolytic, breaking down blood and other tissues. When a human is bitten, the venom causes massive tissue damage that creates a large, persistent, ulcerated wound radiating out from the bite. Infected insect bites and other spider bites are frequently misdiagnosed as brown recluse bites but rarely cause the same level of damage.

The life cycle of spiders and other arachnids is simpler than that of insects. Spiders hatch as miniatures of the adults; they lack larval and pupal stages. Most spiders overwinter as juveniles, but some species overwinter as eggs inside cocoons, and some overwinter as adults. Many species, such as the familiar large black and yellow orb-weaver (*Argiope aurantia*), lay one egg mass and then die, a pattern made famous by E. B. White's classic tale *Charlotte's Web*, but others, such as the equally familiar common house spider (*Achaearanea tepidariorum*), produce more than one egg mass. Enormous nursery web spiders (Pisauridae) guard their eggs and spiderlings until they are large enough to survive on their own (Plate 47). These spiders, which are about 3 1/2 inches long, are usually found near water, and some are known as fishing spiders. Their hairy legs and large area allow them to run over the surface of water. They look similar to the common wolf spiders of the forest.

Daddy longlegs are not spiders (Araneae) but another type of arachnid (Opiliones). They are also called harvestmen because they are common in autumn. Their body is ovoid, with eight extremely long, thin legs, and they move easily over vegetation and the forest floor like all-terrain vehicles. The second pair of legs is mainly sensory and used like antennae (which they lack, along with other arachnids). The creatures tap and probe with them like sensitive fingers, and they also use their shorter, leglike pedipalps near the mouth as sensory probes. They can see through a median eye.

Daddy longlegs do not produce either silk or toxins and do not bite. They are scavengers, attracted to sweet fluids and decaying material. The urban legend that daddy longlegs are poisonous probably results from their confusion with a group of true spiders, the pholcids, which are also called daddy longlegs. (This is a good example of the confusion that can result from common names.) The pholcids build messy webs, often in the dark corners of cellars and caves. They also have very long legs, but their body is shaped like that of a true spider and they have many eyes. Like other spiders, their venom is toxic, but their fangs are too small and weak to bite humans.

## Climatic Conditions in Fall

September in Great Smoky Mountains National Park at Gatlinburg, Tennessee, has an average high temperature of 79°F and an average low temperature of 54°F. An average of 4.6 inches of precipitation falls over seven days. Average day length is 12 1/2 hours. This month can be one of extremes when it comes to precipitation, for it is the most common month for damaging hurricanes. September storms from the Gulf of Mexico often track up the Appalachians, bringing torrential rains and damaging winds. September of 2004 will long be remembered in western North Carolina for hurricanes Frances and Ivan, which hit the same area within a two-week period. Extraordinary rainfall caused rivers and creeks to flood their banks, huge numbers of large trees were downed by the combination of soggy ground and high winds, and mudslides on formerly stable slopes changed the landscape. Such storms, while infrequent, are major causes of ecosystem change. In other years, when hurricanes track farther west, they take with them the major source of precipitation, and drought results. These prolonged dry spells, such as that of 2007–8, also change the landscape by killing plants on ridges and other dry locations. Though more subtle than a hurricane, periodic drought is responsible for the distribution of many plants. As an example, rosebay rhododendrons may climb up dry slopes during a series of wet years, but one drought is enough to reduce their numbers on ridges and open up space for mountain laurel and huckleberry. The fall equinox occurs on September 22 or 23.

October has an average high temperature of 70°F and an average low temperature of 42°F. An average of 3.0 inches of precipitation falls over five days. October is typically the driest month of the year. October 15th is the average first frost date for Gatlinburg, usually brought by the first returning cold front. A cold snap during the first two weeks of October produces spectacular fall leaf color around the middle of the month, but a later first frost may push the leaf color season back into early November. Average day length for October is 11 1/2 hours.

November has an average high temperature of 60°F and an average low of 33°F. An average of 4.0 inches of precipitation falls over eight days. An average of 0.2 inches of snow falls during the month. Average day length is 10 1/2 hours. Several hard freezes occur in November, and this puts an end to most insect activity.

# Cycles of Winter
## December, January, February

After a winter ice storm, whole forests sparkle like diamonds when backlit by the brilliant sun. Water trickling over rock ledges sculpts ice palaces of yard-long icicle columns, and shaded rock faces support massive walls of ice like glaciers meeting the sea. The beautiful forms, however, belie the damage ice can do. For example, formation of ice within cells damages tissue and limits the distribution of many plants and animals. Heavy ice loads can break tree limbs or topple whole trees, leaving a landscape of downed wood. On the other hand, although the trees of the forest may suffer, small animals actually benefit from the fallen wood, which provides food and shelter.

While snow and ice are dramatic indications of winter, more subtle reminders of cold weather can be even more enchanting. On wintry mornings, breath becomes visible, as water vapor condenses in the cold, dry air. In the warmth of summer, I rarely consider such a fundamental aspect of being alive, but in the cold air of winter, I feel like a child who has experienced for the first time the magic of each breath becoming a visible cloud.

## Freeze/Thaw Cycles and the Magic of Ice

From a scientific perspective, the transition between liquid and solid water is equally magical. Water is the only substance on Earth that expands as it freezes, becoming less dense in its solid form. Other chemicals become denser and heavier as they change from a liquid to a solid, shrinking down to occupy less space as they contract, but water does just the opposite. Because of this unusual property, when water freezes and forms as ice, it has several significant effects on the surrounding environment. First, it creates great erosive power. As water freezes its volume increases by 9 percent. This huge increase creates a force of four hundred pounds per square inch, which is roughly equivalent to the bite force of a great white shark. The apparently insignificant but constant drip of water through a crack in a rock face can, upon freezing, dislodge boulders as easily as it bursts the metal pipes under your house. Similarly, if water in the tissue cells of plants and animals freezes, it will kill those tissues, a phenomenon called frostbite in animals. Animals and plants that are adapted to live in cold climates have various ways to prevent their tissues from freezing.

Second, since ice floats on water because it is less dense as a solid than a liquid, the ice on a pond's surface insulates the underlying water from the subfreezing air overhead. Any fluid other than water would freeze, become heavier, and then sink, leaving the exposed liquid surface to freeze and sink again. After several repetitions of this cycle, the entire pond would be frozen solid. Not only would it therefore be extremely slow to thaw, if it ever did (because the ice would stay on the bottom of the pond thawing only from the top as the air warmed), all the plants, fish, and other animals that live in ponds would be frozen and crushed by the solid, sinking ice. Happily, none of this happens because ponds freeze at the surface and the ice floats there, insulating the deeper layers of water from the cold air temperatures. Next time you drop an ice cube in your drink and it floats to the top, remember that you are holding one of the most magical aspects of the natural world right there in your hand!

Expanding ice is often forced up and out of damp soil as the groundwater freezes. As it forms, the ice takes the path of least resistance and squeezes out like toothpaste above the surface of the ground. From heavy clay soils, hairlike fibers of ice emerge and combine into ribbons, creating three-inch-tall crystalline columns (Plate 48). Damp areas along road cuts where clay is exposed are good places to look for these strange ice sculptures. On rare occasions, when an early hard freeze hits while herbaceous plants are still full of sap, these ice ribbons burst from soggy stems like bizarre ice flowers.

## Deciduous versus Evergreen Trees

In the Southern Appalachians, as in most of temperate eastern North America, forests are dominated by broad-leaved deciduous trees. As they prepare to shed their leaves, we are rewarded by beautiful fall color. By contrast, higher Appalachian peaks and the boreal forests of Canada are dominated by evergreen conifers.

FIGURE 5-1
The male and female cones of this conifer, slash pine (*Pinus elliottii*), appear on new growth in early spring. The female cone occurs at the growing tip, or candle, of the branch, and the multiple male pollen cones are grouped at the candle's base.

Broad-leaved trees are angiosperms and produce flowers. The flowers may be showy to attract insect pollinators, such as those of tulip trees, or they may be inconspicuous and wind pollinated, such as those of oaks. The needle-leaved trees, such as pines, hemlocks, spruces, and firs, are gymnosperms, a more ancient group of plants. Instead of flowers, they bear male and female cones (Figure 5-1) and are known as conifers, or "cone bearers." Cycads, gnetophytes, and ginkgo trees are also gymnosperms, but none are native to the Appalachians. (Millions of years ago, some occurred here, and ginkgos have been introduced.) Both angiosperms and conifers may be evergreen or deciduous, but because our forests are mostly either deciduous angiosperms or evergreen conifers, the differences between those two will be the focus here.

Angiosperm leaves are more efficient at photosynthesis than conifer needles. Under similar environmental conditions and over an equal time period, a square inch of angiosperm leaf generally manufactures more carbohydrate than a square inch of conifer needle. There is a price to pay for this efficiency, however. Because deciduous angiosperms devote more of their leaf structure to photosynthetic machinery than to protective tissues, the leaves are more susceptible to herbivore damage and they cannot tolerate freezes. Because the leaves are damaged by freezes and solid ice cannot be transported through the tissues of the plant, deciduous trees prepare for winter by shutting down photosynthesis and dropping their leaves. Replacing leaves in the spring requires stored energy (sunlight), water, and nutrients from the soil, and the more often leaves are replaced, the more energy, water, and nutrients are used. Thus deciduous angiosperms dominate areas where there is a long season of sunlight, adequate soil water, and good soil nutrition but where winter freezes damage their delicate leaves.

If the deciduous angiosperms are the hares of temperate forests, then evergreen conifers are the tortoises. Infertile soils favor evergreen leaves because an evergreen tree uses fewer nutrients over its lifetime in manufacturing leaves. Conifers are also predominant in dry habitats because they are better at conserving water. Their leaves are coated with a waxy cuticle, and their slow rate of photosynthesis means less water is lost through transpiration. Finally, evergreen leaves are advantageous in habitats with a short photosynthetic season. As soon as the soil thaws, evergreens can begin photosynthesis, in contrast to deciduous trees, which must first re-grow their leaves before they can begin to manufacture food through photosynthesis. In addition, the shape of a conifer tree makes it more effective at shedding snow than the typical broad-leaved tree. Thus evergreen conifers tend to dominate in areas with short photosynthetic seasons or infertile or dry soils, such as our high mountain peaks.

The evergreen leaves of conifers are adapted to meet these stringent environmental challenges. Conifer needles, as well as other tissues of the tree, are full of a sugary and thick, strong-smelling resin. This resin acts as an antifreeze, lowering the freezing point, thereby protecting the tissues from damage by cold. As an added benefit, the resins also discourage feeding by insects. At the end of the growing season, the deciduous leaves of tulip trees are so full of holes that it is nearly an impossible task to find a perfect one among the fallen leaves. Pine needles, on the other hand, are rarely damaged. Needles that remain on the tree all winter serve to break up the flow of air across the branches, reducing evaporation and heat loss from wind chill.

Even though conifers are evergreen, they still shed old leaves and grow new ones as part of a general growth cycle, but as noted earlier, this growth cycle is much longer than the annual cycle of deciduous trees. Because they shed leaves that are full of resin, resistant to both rot and insect damage, the forest floor under large conifers is often spongy and soft from the accumulated layers of dead needles. These needles act as effective mulch, slow to decay and limiting the number of other plants that can gain a toehold.

## Appalachian Conifers

In the Appalachians, the shortest growing season and poorest soils are found on the mountain peaks, which are dominated by spruce-fir forest. The highest peaks in the southern mountains host Fraser fir (*Abies fraseri*), with balsam fir (*A. balsamea*) taking over to the north. Firs are tolerant of shade and retain leaves on their lower branches, giving them the perfect Christmas-tree shape, and among the conifers, they are most beloved for that use. Fir trees can be identified by soft, flat foliage and cones that project upward from the limbs. Their inch-long needles have two white stripes on their underside and produce a pleasant smelling resin when they are stroked. This resin, known commercially as Canada Balsam, has been used for many years to glue specimens to microscope slides because its refractive index is the same as that of glass.

Red spruce (*Picea rubens*) is the only spruce native to the southern mountains and occurs with Fraser fir on mountain peaks. It has stiff, sharp-pointed needles with four sides, and its cones hang down from the limbs. The needles will prick you if you run your hand along the limbs. Old-timers call these trees "he-balsams" in contrast to soft fir trees, which they call "she-balsams." She-balsams (firs) also drip milky resin from wounds. Several spruces, including the non-native, drooping limbed Norway spruce (*P. abies*), are often planted horticulturally.

Eastern hemlock (*Tsuga canadensis*) is an evergreen conifer that is water-loving and shade tolerant. It grows in low-elevation mountain coves and ravines where it often forms shady stands along streambeds. Its slow but steady photosynthetic rate allows it to compete successfully in these moist, shady habitats. Hemlocks are long-lived, with some trees living as long as eight hundred years. The low, spreading limbs are adapted to shed snow, and the trees are rarely damaged by it. Their tiny cones hang pendulously from the branches. The half-inch-long, flattened leaves are soft and silvery, occur in a single plane on opposite sides of the stem, and have two white stripes on their underside. Since the leaves of Fraser firs and eastern hemlocks are so similar, the trees can be hard to tell apart. Hemlocks, however, have shorter leaves, and their branch tips droop. Furthermore, the habitats of fir and hemlock rarely overlap. The rarer Carolina hemlock (*T. caroliniana*) occurs on rocky ridges. It has larger cones than its relative, and its leaves occur all around the branches instead of in one plane.

An introduced aphidlike pest, the hemlock woolly adelgid (*Adelges tsugae*), threatens these graceful and ecologically important native trees. It appeared in Virginia in the 1950s after being accidentally introduced into the Pacific Northwest in 1924 from Asia. Severe winters appear to control its spread into the Northern Appalachians, but the hemlocks of the Southern Appalachians have not been spared. Eighty percent of the hemlocks in Virginia's Shenandoah National Park have already died. The adelgids were first found in the Great Smoky Mountains National Park in 2002. Pesticides and insecticidal oils are being used on a limited basis. Predatory beetles (*Sasajicymnus tsugae* and *Laricobius nigrinus*) have been released as biological controls more widely, but whether the beetles can reduce the adelgid infestation remains uncertain at this point. Another beetle, the introduced Asian multicolored ladybug (*Harmonia axyridis*) is also a predator of the adelgids (Plate 49), but only on trees in sunny locations. Since the most ecologically important trees grow along shady streams, they are not helped by these beetles. Many hemlocks in the southern mountains are already dead (Plate 50).

Eastern white pine (*Pinus strobus*) is the most common pine in the Appalachians, although several others occur. Widespread in eastern North America, it is the only pine whose leaves occur in bundles of five (count W-H-I-T-E) instead of two or three. It produces a whorl of branches each year, and the number of whorls provides an estimate of the tree's age. On old trees, the lower branches self-prune, and on northern trees that have been frequently topped by ice storms, the whorls are usually so close together they are obscured. Under ideal conditions, white pines get as tall as two hundred feet. They were the primary source of ship masts during colonial times, and trees emblazoned with the king's cross were earmarked for that use only.

## The "Perfect Storm" of Acids, Adelgids, and Global Warming

On the highest peaks of the Appalachians, the evergreen trees of the spruce-fir forest are dying. The Great Smoky Mountains National Park (GSMNP) and Mt. Mitchell, the highest peak in eastern North America, have been particularly hard hit. Several factors have combined to cause tree kill. The first is acid rain, which is defined as rainfall with a pH lower than 5.6. (On the pH scale neutral is a 7. Lower numbers on the scale indicate acidity.) Typical rainfall in GSMNP today has a pH of 4.5. Because each unit on the pH scale represents a factor of ten, that means rain in the park is ten times more acidic than normal rain. And more damaging than acid rain is acid fog. Acid fogs are commonly recorded with a pH of 2.0, making them about 5,000 times more acidic than normal rainfall, equivalent to the acidity of vinegar.

Can you imagine the effect of watering your houseplants with vinegar? Fogs linger, drenching the foliage and soils with water and the dissolved acids it contains. Acids are even deposited as dry, dustlike particles. To compound the problem, the acids in the water cause toxic metals such as aluminum to be released from the soil while beneficial nutrients are immobilized. Thus, the trees are being doused with acid, poisoned by soil chemicals, and starved of nutrients.

The acids, mainly compounds of sulfur and nitrogen, are a byproduct of fossil fuels, which are burned to provide energy for industry and vehicles. GSMNP receives the prevailing winds from the industrialized Ohio River valley, and the high peaks trap this chemical-laden air. In addition, the park itself is the most heavily visited of all U.S. National Parks, receiving over 9 million visitors yearly. Most of them arrive in automobiles, and their vehicle exhaust contributes to the problem. Ozone, another byproduct of fossil fuel use, is also damaging to plants and animals because it is a highly reactive form of oxygen. Ozone in GSMNP during summertime, when natural conditions favor its formation and visitation rates are highest, frequently reaches damaging levels.

Air pollution is not the only problem facing the spruce-fir forest. The introduced balsam woolly adelgid (*Adelges piceae*), an aphidlike pest that was first recorded in the Southeast in 1956, is another threat. It was introduced accidentally from Europe into the northeastern United States around 1900 but was not responsible for severe damage until it reached the pollution-weakened firs of the Southern Appalachians. Over 95 percent of the Fraser firs in GSMNP have already died. There is some hope that the ecosystem may recover because young trees are least affected by adelgids, but recovery can only begin once fossil fuel use is phased out. Unfortunately, the increasing population of white-tailed deer, who feed on the young trees before they reach reproductive age, is threatening the natural seed stock. Park staff members are currently collecting and storing seeds and seedlings for the future.

A final affront to the high-elevation spruce-fir forest is global warming. As climate warms, species that are adapted to colder temperatures must migrate toward the poles or to higher elevations. Because the spruce-fir forest is already clinging precariously to the highest elevations in the Southern Appalachians, it has nowhere else to go.

Introduced species are a major cause of current extinctions, second only to habitat loss. An invasive species is defined as an alien species whose introduction causes harm. Not all introduced species cause harm, but those that outcompete, displace, or weaken native species compromise the stability and overall health of native ecosystems. Invasive species may be plants, animals, fungi, bacteria, or protists. Some are pathogens, such as chestnut blight (see chapter 4), or parasites, such as adelgids. Others are predators like feral cats, which reduce native songbird populations (see chapter 2). Many species of invasive plants, such as Chinese privet (*Ligustrum sinense*) displace native plants from their natural habitat.

By definition, a native species cannot be an invasive. However, some native species have increased their numbers at the expense of other species, and their populations have grown to such levels that they have become pests. White-tailed deer fall into this category. Before about 1950, deer were rarely encountered throughout much of the East, but as more land opened up and their predators were eliminated, their numbers grew. Now, some herbaceous wildflowers have been eliminated from areas of heavy deer browsing, and deer have been implicated in the reduction of forest diversity and density.

## American Holly: A Broad-Leaved Evergreen Tree

American holly (*Ilex opaca*) is one of the few evergreen flowering trees in the Appalachians. Its small, greenish yellow flowers are inconspicuous in the spring, but the bright red berries produced by female trees stand out against the dark green leaves all winter. These berries, and the seeds they contain, are generally not consumed and dispersed by birds until late winter or early spring of the year after they flowered. The berries contain unpalatable compounds that degrade over time. By winter's end, when there are few other berries left to consume, cedar waxwings and other birds devour holly berries.

All hollies, even the many deciduous species, have small spines on the edges of their leaves, but in the American holly, these spines are long and rigid. The fibrous and tough leaves are also covered with a waxy cuticle and filled with chemicals that protect the leaves from damage by cold temperatures and make them unpalatable to herbivores. Few leaves of American holly are grazed by animals such as white-tailed deer or caterpillars, at least until the leaves age and, like the berries, become more palatable. Another introduced

FIGURE 5-2
Leaves of *Rhododendron maximum*
droop and curl tightly at 20°F.

insect, the two-banded Japanese weevil (*Pseudocneorhinus bifasciatus*), seriously damages the leaves of American holly and many other plants.

## Rhododendrons as Evergreen Thermometers

Several species of *Rhododendron* and mountain laurel retain their leaves during winter. As temperatures drop below freezing, their leaves droop and curl. It's possible to estimate the air temperature from the degree of leaf curl. The plant most often used as a thermometer is rosebay rhododendron or great laurel (*R. maximum*).

The leaves respond to cold by first drooping, then curling lengthwise. At temperatures above freezing, the flat-bladed leaves extend at an approximately right angle from the stems. Just below freezing, the leaves droop downward but remain flat. At around 25°F, the leaves begin to curl, and by 20°F, they are curled so tightly that they look like thick green pencils (Figure 5-2). Once the temperature drops below 20°F, they can't curl up much tighter, and only a real thermometer will tell you how cold it is.

The immediate cause of the leaf-curling response in *Rhododendron* is unclear, but we do know that loss of water from the leaf (wilting) is not responsible, although water may be displaced or change form within the leaf. If you place a freshly picked leaf in a freezer, it will curl in minutes but then recover its normal shape when allowed to warm at room temperature. On the plant, drooped and curled leaves expose less of their surface area to sunlight and this fact has prompted an adaptive explanation. At subfreezing temperatures, the photosynthetic machinery of the leaf is more susceptible to damage from light than at warmer temperatures. For this reason, leaf curling and drooping under cold conditions may help the plant to avoid such injury.

## Evergreen Herbaceous Plants: Cycles Reversed

Some of our terrestrial orchids, such as puttyroot (*Aplectrum hyemale*) and cranefly orchid (*Tipularia discolor*), are more noticeable in the wintertime, when their four-inch-long leaves are welcome splashes of green against the background of dead, brown leaves on the forest floor. In both species, the underside of the leaf is darker in color, and in cranefly orchid, it is purple. This darker color may reflect light back through the photosynthetic part of the leaf, or it may absorb the light that would otherwise pass out, thus warming the active part of the leaf. The leaves die back in late spring, when small, brownish flowers form on a stalk about a foot in height.

This reversed seasonal cycle of leaf growth is related to light availability. The deciduous trees of the rich forests where these orchids grow intercept most of the light from mid-spring to mid-fall. These orchids have leaves when the trees do not, so they gather light and manufacturing carbohydrates in late fall, early spring, and warm winter days when the soil is thawed. They save that energy in large, underground stems called corms and use it to flower. Like the spring ephemerals (see chapter 2), they gather light energy in the "off-season" before the trees intercept it all, but the orchids have the whole of the winter season and late fall as well as early spring to harvest sunlight.

A few other herbaceous plants are evergreen. In addition to the leaves of galax and Oconee bells (see chapter 2), those of rattlesnake plantain (*Goodyera pubescens*) are a welcome sight year-round. The rosettes of two-inch-long leaves are green, but their veins are mottled with white. Because the variegated leaves resembled snakeskin, it was once believed that the plant provided an antidote to snake venom. Rattlesnake plantain is the most common Appalachian orchid. In summer it sends up a foot-high stalk that bears a cluster of small white flowers.

Spotted wintergreen, or pipsissewa (*Chimaphila maculata*), is an evergreen heath whose leaves are marked somewhat like those of rattlesnake plantain. They are dark green with a white region along the midrib. The small plant is usually less than six inches high. It blooms during the summer months, when a three-inch stalk emerges from the whorl of uppermost leaves to bear a pair of waxy white flowers. This woody, persistent tiny shrub of the forest is collected for its medicinal properties but remains common. As a child, I imagined that these engaging little plants were really giant trees and created a whole fantasy world of plants, animals, and people that lived in the shade beneath them.

Trailing arbutus (*Epigaea repens*), like pipsissewa, is among the smallest of the evergreen heaths. I see it most commonly along road banks, where its hairy stems and two-inch-long evergreen leaves trail over the ground. The small, pale pink or white flowers bloom in early spring, and bending over, or even lying down, in order to catch their delightful scent is worth the effort.

Perhaps my favorite evergreen plant is the humble partridge-berry (*Mitchella repens*). The little plant grows prostrate on the ground, its numerous stems and small, rounded leaves in opposite pairs creeping gently through open woods. In early summer, the small, white, four-petaled flowers bloom in pairs (Figure 5-3) that are fused at the base. Later in summer, berries form, which persist into winter and, as the plant's name suggests, are eaten by partridges (ruffed grouse) and other birds. Because each pair of flowers forms a single berry, partridge-berry was added to the bouquets of mountain brides as a reminder that a good marriage is the joining of two into one.

## Primitive Plants and Their Reproductive Cycles

Mosses, clubmosses, and ferns are not flowering plants, or even seed-producing plants. They are primitive land plants (mosses are the most primitive), the first true plants (more complex than algae) that invaded land about 400 million years ago. Instead of seeds, they produce tiny spores. These early plants dominated the landscape for millions of years, producing forests that looked very different from those of today. During the Pennsylvanian period, about 300 million years ago, forests were composed of tree-sized ferns, horsetails, and clubmosses. By the time of the dinosaurs, 150 to 65 million years ago, gymnosperms were dominant and were the main food source for the giant herbivorous reptiles. Forests full of flowering trees appeared in the

FIGURE 5-3

The twin flowers of partridge-berry (*Mitchella repens*) are united at the base, and each pair of flowers eventually forms a single berry.

last 70 million years or so. Our current forests contain relicts of those earlier times, with ferns and clubmosses still represented but as smaller and less dominant plants.

Christmas fern (*Polystichum acrostichoides*) is the most common evergreen fern in the Southern Appalachians. Its common name is derived from at least two sources: it is green at Christmas time, and its cut fronds often supplied decorative greenery for early settlers during the Christmas season. In addition, each small, inch-long leaflet is shaped like a Christmas stocking. The large fronds are composed of about twenty pairs of these little boots lined up along both sides of the central axis of the leaf, like stockings hung from a mantel, but these stockings are hung by the heel.

In winter, some of the tips of the large, foot-long fronds of Christmas ferns are dark and shriveled. These dried leaflets bore the fern's reproductive spores, and once the spores were released from the plants, the leaflets died back. In the spring and summer, when the entire frond is alive and healthy, the undersides of the living, terminal leaflets bear small, brown packets of spores. Different ferns produce their spore packets in different positions—

Fossil fuels do, indeed, contain fossils. These fuels are nonrenewable because they were formed from particular plants that are no longer dominant in our forests. Most of the enormous coal beds of the Appalachians were formed during the Pennsylvanian period (300 million years ago). Primitive, nonwoody plants grew in swampy locations and were transformed into beds of coal when those swamps were covered by sediment and then subjected to heat and pressure. Fossil fuels are composed mainly of carbon and hydrogen (hydrocarbons), on which all life is based. When fossil fuels are burned by adding oxygen, carbon dioxide is the primary by-product. When we burn fossil fuels, we release the photosynthetic energy that has been stored in those plant bodies for millions of years. The sulfur in most coal, which when combined with oxygen forms the acid pollutant sulfur dioxide, was formed by the bacteria that lived in those same swamps. Such bacteria are still present today in brackish marshes.

sometimes on special fronds, sometimes only on leaflets in the center of the frond, and sometimes at the tip of the frond.

Unlike ferns, nearly all Appalachian clubmosses are evergreen. Most are less than six inches tall. Their leaves are small and scalelike, borne on upright stems. Frequently the stems travel horizontally across the ground, so you'll find large patches of clubmoss on the forest floor.

Most clubmosses reproduce during fall and early winter. Turkey brush (*Diphasiastrum digitatum*), also known as running cedar or ground-pine, sends up a stalk with three upright sections, reminiscent of Neptune's trident or a diminutive candelabrum. Common ground-pine (*Dendrolycopodium obscurum*) looks rather similar, but the spores are held in a single-sectioned stalk (Figure 5-4). Shining clubmoss (*Huperzia lucidula*) produces its yellow spores at the bases of the leaflets. If you touch these parts of the plant, or the wind blows through them, puffs of spores are released. The plants are prolific, releasing a dense yellow cloud of spores from each structure, and a large area covered by the plants can produce a huge number of the spores. Historically, the spores were collected for the medical industry to coat aspirin and other pills to prevent them from sticking to each other. The spores were also used in early photography as a component of flash powder. Today, clubmosses are used in homeopathic remedies and have been suggested to cure everything from diarrhea to impotence! A Chinese species is being investigated as treatment for Alzheimer's disease.

FIGURE 5-4

Clubmosses are evergreen and produce spores in fall and winter. The spores of common ground-pine (*Dendrolycopodium obscurum*) are held in packets above the leafy stem.

## Lichens and Jelly Fungi

The winter sky, a dome as blue as a robin's egg, seems supported by the sturdy, steel gray columns of trees underneath it. Those seemingly lifeless columns, however, are covered in chalky green swatches, bumpy crusts, and fishlike scales called lichens. These plantlike creatures eke out an existence in extreme environments because they are composites of two different life forms that cooperate rather than compete.

Lichens are formed by the symbiosis between a fungus and a single-celled photosynthetic alga (or, less commonly, a bacterium). The alga is provided with a protective body and the fungus with food. The resultant new form of life is greater than the simple sum of its partners' parts. A lichen can grow where neither a fungus nor an alga can grow—on rocks, in direct sun, and high up on trees where either symbiont would dry out and die. In fact, lichens are often the first type of life to colonize bare rocks, paving the way for plants to follow. Algae, in contrast, usually live only where they stay constantly moist, and fungi grow where they have a steady food supply of decomposing material, usually underground or within another organism.

In order to achieve this mutually beneficial lichen symbiosis, the partners have nearly given up their individual identities. If the two are separated from each other, each can grow individually, but in a form that differs so completely from the lichen's shape that it is unrecognizable as a component of the lichen. Once split, the fungus and alga cannot be recombined under laboratory conditions; only nature produces lichens.

There are three basic categories of lichens but many variations within each. The crustose forms grow like a thin film or crust and adhere so tightly to a tree or rock that to remove them would require scraping them into pieces. The foliose, or leaflike forms, are flattened and lift up from the surface on which they are growing. Finally, the fruticose, or shrubby, forms hang suspended from trees like long gray beards; they are often used in model train sets as miniature shrubs.

Leaflike lichens, such as ruffled lichens (*Parmotrema* and *Rimelia* spp.), are common on tree bark. Their edges usually curl up, revealing their lower brown or black surface, and are fringed with tiny black hairs, like eyelashes. They do not damage the trees on which they grow. Rock tripe (*Umbilicaria mammulata* and *U. muehlenbergii*), gray disks several inches in diameter, are dark underneath and adhere to rock with the center of the disk. Most lichens are edible, if not tasty. Boiled rock tripe reportedly tastes like tripe, the stomach of cows and sheep. It is, however, slow growing and cannot be recommended for harvesting.

British soldiers (*Cladonia cristatella*), which grow on tree stumps and on the ground are among the best-known shrubby lichens. They are named for the red top that is perched on each inch-tall, fingerlike projection like the red hat of a British soldier. There are several closely related species in the Southern Appalachians, including *C. floerkeana*. Several species of reindeer moss (*Cladina* spp.) are common shrubby lichens that are recognized by many people; in the Arctic they are an important food source for caribou and reindeer. Lichens are also a major food source for northern flying squirrels. Beard lichens (*Usnea* spp.) dangle from trees like untrimmed goatees. As reflected in their scientific name, they are similar in appearance to Spanish moss (*Tillandsia usneoides*), a bromeliad related to pineapples that grows along the coastal plain of the Southeast.

In contrast to lichens, which are conspicuous in wintertime, most fungi are dormant. One of the few common winter fungi is tree ears (*Auricularia auricular-judae*), a jelly fungus found throughout North America. The brown, rubbery mushrooms frequently cover the dead branches of trees on which lichens are also present. They can freeze and thaw several times without

Because lichens do not have an impermeable cuticle like most plants and animals, they are not able to prevent pollutants from entering their body. Anything that touches them is absorbed. Fungi, the main component of the lichen body, naturally absorb nutrients as well as pollutants. Since lichens are present year-round, unlike the leaves of deciduous plants, they are constantly exposed to air pollution. If they're exposed to enough air pollution, they'll die. In general, shrubby lichens are most susceptible because they are exposed in three dimensions, while crustose forms are least susceptible. Localities that are host to many different types of lichens generally have good air quality.

---

damage to their tissues. Tree ears are edible, and a closely related mushroom species is grown commercially for use in Asian-style soups and other foods.

## Springtails: Enigmatic Winter Animals

On occasional winter days, I have been walking through the forest and been surprised to see hundreds of tiny green or black dots crawling over the snow or leaves. If a recent rain has left puddles in the leaves, dozens of the tiny creatures can be found floating on the surface tension of each puddle. If you look at them with a magnifying glass, you will see that they are wingless, six-legged insects with a pair of prominent antennae (Figure 5-5). They are springtails or snow fleas (Collembola), denizens of the leaf litter. Some species (not snow fleas) do indeed have tiny springs on their rear ends, and they hop about like fleas when disturbed.

What drives them above ground in the middle of winter? Few other insects are present in winter because insects, like reptiles and amphibians, rely on their environment to provide them with enough warmth to operate. Midges and other aquatic nymphs occasionally emerge from their relatively warm water sources during the winter season, and a few furry brown moths beat themselves against my lighted windows, but most insects are absent during the winter months. How the tiny springtails manage to find enough heat to crawl over snow is a mystery.

Springtails are primitive insects. A 400-million-year-old fossilized springtail (*Rhyniognatha hirsti*) is considered the oldest known insect. A 428-million-year-old fossilized millipede (see chapter 3) is the oldest land animal on record; millipedes are taxonomically distinct from insects.

FIGURE 5-5

Springtails or snow fleas (Collembola) are tiny, primitive insects that live in leaf litter. They are wingless and have six legs and a pair of antennae. This fairly common species, *Thalassaphorura encarpata*, is a beautiful pale green color.

## Hibernation and Denning Cycles in Groundhogs and Bears

Because winter is the season of cold temperatures coupled with scarce food, some animals spend it in hibernation (see chapter 1). The only mammal that truly hibernates in the Appalachians is the groundhog (*Marmota monax*), also known as the woodchuck. During hibernation, its heartbeat drops from nearly one hundred beats per minute to four, breathing rate drops to about one breath every three to four minutes, and body temperature decreases from 98°F to around 40°F, the temperature of the burrow. Hibernation begins in early November and lasts until early March, as determined by day length. Animals that are kept at constant day length do not hibernate.

Short and squat with brown fur, groundhogs are actually ground squirrels, although certainly the biggest squirrel in the region. In fall, adults can weigh up to twelve pounds, as they fatten up in preparation for hibernation. Like other rodents, they eat vegetation. They prefer to live where grasses are plentiful, and they usually construct their burrows on the borders of open fields and along roadways. The expansion of roads, agriculture, and suburban woodlots has increased the population of these grass- and shrub-eating mammals, and their numbers, like those of white-tailed deer, are greater now

According to the city fathers of Punxsutawney, Pennsylvania, a groundhog named Phil predicts the six weeks of weather following Groundhog Day on February 2nd. If it is sunny, so that Phil sees his shadow and runs back into the burrow, then winter is expected to last another six weeks. Most of the time (90 percent, according to the official website), Phil sees his shadow and six more weeks of winter weather are forecast. Is it surprising that nine out of ten Pennsylvania winters last into mid-March? New England farmers have a saying, "Groundhog Day—Half Your Hay," meaning that only half the winter is gone in early February, no matter what a ground squirrel might predict; if more than half your hay is gone by then, your farm animals are in trouble. Several English songs and Scottish rhymes make connections between current and future weather patterns, and the European hedgehog was used for weather prognostication. Groundhogs look something like hedgehogs, so when Germans settled in Pennsylvania, they naturally turned to the groundhog to make their weather predictions. Ironically, hedgehogs are insectivores, more closely related to shrews and moles than to the ground squirrels, which are rodents. They do, however, hibernate. When the official Pennsylvania celebration began in 1887, participants dug a hibernating groundhog out of its burrow and waited for the disturbance to awaken it, but now Phil has an artificially warmed and lighted burrow to ensure he is awake for the service.

than when Europeans arrived. Because they eat vegetation, they can become pests around gardens and agricultural fields.

Groundhogs dig burrows in which they sleep and hibernate. The opening is often conspicuous because they mound up the excavated dirt around the entrance. Sometimes the opening is exposed, but it is often located under brush or high weeds, which provide some protection. The burrows may have several well-concealed side entrances for emergency use. Groundhogs are prey for many of the carnivorous mammals as well as large raptors. Humans also kill and eat them. Whole families may live in the burrow system, which is composed of several nest chambers lined with grass and leaves, but males and unmated young adults are often solitary.

Black bears (*Ursus americanus*) do not hibernate in winter dens in the same fashion as groundhogs. Their body temperature and metabolism drop only slightly and they awaken frequently and easily—as some bear researchers have unwittingly discovered! However, they do not defecate or urinate for the entire time they remain in their den, a phenomenon that remains a

If you live in black bear territory, minimize your effect on bears by properly storing garbage, pet food, bird feeders, and outdoor grills where bears cannot get to them. Under no circumstance should you feed bears; if you do, they will associate humans with food. It is easier to prevent a problem from occurring than to attempt to solve it afterward. A nuisance bear cannot be relocated because there is nowhere to put it that is out of human contact. In addition, new bears will simply move into the place from which the nuisance bear was removed. The only viable solution to a persistent bear is to kill it. While fatal attacks on humans are rare, they have occurred in the Southern Appalachians.

puzzle for scientists. The young are born while the mother is denning, and she nurses them for several weeks before she emerges from the den to feed. Human-fed black bears den later in the season than those under natural conditions, so it is likely that they enter their dens when food sources decline.

Black bears formerly occupied a wide range over the eastern states but are now restricted to the mountains of the Appalachians and coastal swamps of eastern states, where there are large areas of undisturbed forest. Within these restricted ranges, however, the number of bears has increased in recent years. For example, in North Carolina, the bear population almost tripled between 1971 and 2004, from about 4,000 to approximately 11,000. As the number of bears and humans increases and as more humans move into bear territory, encounters with black bears are becoming more common. In 1993, the North Carolina Wildlife Resources Commission received 53 complaints about nuisance bears, but in 2007, they received 418.

Black bears' diet also contributes to their conflict with humans. Black bears are omnivores, eating many different kinds of foods, from grasses, berries, and acorns to grubs, insect larvae, and carrion. They will raid trashcans, bird feeders, and areas where pet food is stored. (See sidebar "Help Black Bears and People.") They have an excellent sense of smell, so they can locate stored food, and they are large enough to cause damage getting to it. I have had a couple of close encounters with black bears. One visited my bird feeder, and another (or the same one on a different date) raided a metal trash can in which we stored trout food. The bear left a telltale hole the size of the diameter of a pencil in the metal lid. Researchers who track black bears look for such holes, which bears make with their canines, in the sardine cans they use as bait. Only a black bear has the strength and the canines to pierce the metal can. Male bears typically weigh between 130 and 300 pounds, but they can weigh in excess of 600 pounds; females typically weigh between 90 and 150 pounds.

## Activity Cycles of Small Mammals: Shrews, Moles, Mice, and Flying Squirrels

Some mammals are too small to hibernate. Shrews, moles, mice, and flying squirrels, for instance, are unable to store enough fat to sustain them for extended periods of hibernation and fasting. Because they are small, they have a higher metabolic rate than large animals. Their small bodies use more energy, ounce for ounce, because they have a small body volume to generate heat but a large surface area of skin over which that heat is lost. With a high metabolic rate and poor ability to store food as fat, small animals such as shrews eat more or less constantly.

When you are at rest, your heart beats an average of sixty times each minute and goes unnoticed by all but the most attentive. After exercise or a good scare, your heartbeat can increase to 120 beats per minute, and the throbbing grabs your attention. Now imagine a heart rate of 1,200 beats per minute, 20 beats each second, and you have entered the high-speed world of a shrew, one of earth's most frenetic animals. If we had a shrew's metabolism, sixty times faster than our own, we would be bundles of boundless energy, zooming rapidly from place to place. We would be hungry all the time, needing constant fuel to maintain such a blistering metabolism. And we would die more quickly, for an animal's average heart rate is related to life span: the higher the rate, the shorter the life. Instead of a seventy-year life span, we would live but a year in a state of wild abandon. We would be untamed shrews.

Shrews are insectivores, but they also eat earthworms and snails. They may occasionally eat oily seeds or nuts. Their metabolism is so high that they could not live on vegetation alone. Those kept in captivity eat three times their own body weight every day and will starve to death if deprived of food overnight. Their numerous sharp, pointed teeth reveal this dietary preference. Their saliva is poisonous; when they bite the prey, saliva on their teeth enters the wound and paralyzes the animal.

Shrews and moles (Insectivora) are related to the insectivorous bats (Chiroptera). Not only are their diets similar, but some also use echolocation to find their prey. Shrews tunnel under leaf litter in dark runways. Instead of relying on eyesight in these dark tunnels, they use hearing to detect the reflections of their high-pitched squeaks that bounce back from objects in front of them. Their eyes are relatively small because sight is a less important sense for them.

One of the most commonly encountered shrews in the Appalachians is the northern short-tailed shrew (*Blarina brevicauda*). Including its tail, which is one-third its length, it is over five inches long. In most shrews (*Sorex* sp.) the tail is more than half the length of the body. The short, dense fur of the northern short-tailed shrew is gray to black. Its metabolism is a little less frenetic than the least shrew (760 heartbeats per minute versus 1,200).

The eastern mole (*Scalopus aquaticus*) is the most common mole of the Appalachian region. Larger than the shrew, at over seven inches long, it has very large front feet adapted for digging, and eyes that are permanently closed by flaps of skin. Like shrews, moles are rarely seen, but their tunnels are conspicuous in lawns. Whereas the shrews tunnel through leaf litter, the moles actually dig in the soil. Although moles may dig up your lawn, they do not eat flower bulbs. Rather, rodents such as mice or meadow voles (see chapter 1) do the damage with their large, chisel-like incisors. If you find a dead mole in your yard, you have lost an important control of pesky insects. They are especially effective against the grubs of beetles such as Japanese beetles, exotic pests that are hard to manage. Two other moles also occur in the Appalachians and are easy to distinguish from the ubiquitous eastern mole. The hairy-tailed mole (*Parascalops breweri*) does, indeed, have a thick, hairy tail, and the star-nosed mole (*Condylura cristata*) has fleshy appendages radiating out from around its nose.

Like shrews and moles, mice are too small to hibernate and remain active year-round. Mice are arguably the dominant prey item for many of our predatory animals, especially foxes, weasels, owls and hawks, and snakes. Since the mice population is faced with significant loss to predators, mice are phenomenally prolific. Each year they produce three or more litters of up to seven young, which themselves can become reproductive within one month after birth. The most common mouse in the Southern Appalachians is the deer mouse (*Peromyscus maniculatus*), although other species, such as the white-footed mouse (*P. leucopus*), are also abundant. The deer mouse (Figure 5-6) is named for its distinctly bicolored tail, which is brown on the top and white on the underside, like that of a white-tailed deer.

Flying squirrels are also small mammals that are active year-round. They are nocturnal, only emerging after darkness, and most people never see them for that reason. They are notorious, however, for finding their way inside

FIGURE 5-6
Mice, like the deer mouse (*Peromyscus maniculatus*) pictured here, are active year-round and are an important prey species for most carnivores.

houses, bird boxes, and chimneys, allowing an occasional close-up view. Up close, they are adorable! Because they are nocturnal, they have very large, dark eyes, the better to see with in dim light. Their ears are also larger and more prominent than a gray squirrel's. Their soft, silky fur is an attractive gray on the back and white on the belly, with a dark border where the back and belly fur meet. They are about half the length of a gray squirrel and about one-fifth the weight.

Flying squirrels don't really fly but glide from tree to tree, aided by furry flaps of skin that extend along each side of the body between the front wrist and rear ankle. Their tail is flattened and acts as a rudder during this gliding flight, allowing them to veer and bank to avoid obstacles in their path. When landing on a tree trunk, they swoop upward so that their back feet hit first and their head points upward, allowing them to scamper quickly to the top of the tree in preparation for another glide. They must always start off from a high position and glide to a lower one.

There are two species of flying squirrel in the Appalachians, the southern and the northern. Southern flying squirrels (*Glaucomys volans*) prefer to eat acorns and seeds and are regular visitors to bird feeders. Perhaps because they have historically had little contact with humans, these squirrels are un-

If most of the food disappears from your bird feeder at night, suspect the southern flying squirrel and mice. Nocturnal (and diurnal) mammals are the beneficiaries of much of that food that is intended for the birds! Gray squirrels are notorious raiders of bird feeders. Raccoons and black bears are also attracted to bird feeders and are large enough to cause some damage to them. Minimizing contact with raccoons is important because they are the most common carrier of rabies. Minimizing contact with black bears is prudent because they are large enough to be dangerous and to cause damage to structures around your home. If you feed birds, suspend the feeders so that larger animals cannot reach them, and if you suspect a large animal is helping itself, stop feeding immediately so that it moves on. The last thing you want to do is reward a raccoon or a black bear with food after it has caused damage to get to it!

afraid and rarely aggressive. They are little disturbed by light or proximity of people who are quiet and calm, and will continue to feed even if someone is standing quietly within a few feet of the feeder with a lantern. Because they are active year-round, they visit feeders any time their natural food supplies are low and often congregate at feeders on cool winter nights. They are common animals as well as social, nesting together in hollow trees, attics, or bird boxes during the winter season as a way to keep warm. Once a food source is discovered, the members of the den will visit it and carry away food to store for cold periods when they do not leave their den.

Northern flying squirrels (*G. sabrinus*) are a different story. They occur in Canada and the northern United States, with isolated, relict populations in the Southern Appalachians. These relict populations are restricted to high-altitude spruce-fir forests mostly above 5,000 feet and are endangered. Rather than seeds, northern flying squirrels eat lichens and fungi. Although larger and darker in color than the southern flying squirrel, the northern is difficult to distinguish from its cousin unless in hand.

## Common Nocturnal Omnivores: Raccoons, Skunks, and Opossums

Winter can be a lean season for animals that remain active year-round. Since less food is generally available, these animals often congregate around concentrated sources of food, such as bird feeders or garbage cans. Some may scavenge, but others are attracted by the small prey animals that are themselves attracted by the feeders. Their bold foraging and the seasonal lack of cover make them easier to see in winter.

Among the legions of animals flocking to "bird" feeders are raccoons (*Procyon lotor*), which climb the feeder poles at night like overstuffed cats. The black bandit mask and ringed tail make raccoons among the mammals easiest to identify. Because their back legs are longer than the front, they tend to look hunched up and ungainly. They can, however, run surprisingly fast and are adept at climbing trees, especially when threatened. They are common over most of the United States.

Raccoons have adapted to humans and live in suburbs as well as forests. They nest in hollow trees, in underground dens, or even under buildings. They mainly eat plant material, but they also relish crawfish and frequently scavenge, overturning compost buckets or recycling cans. Their sensitive noses direct them to potential food sources, and their nimble fingers open any container. They can reach twenty pounds or more in weight.

Although it is commonly reported that raccoons wash their food before eating, they are not truly so fastidious. Since their fingers are very sensitive, and this sensitivity is enhanced by water, they manipulate their food in water to feel it better. Raccoon paw prints are often seen in muddy areas near water. They look like tiny, slender human handprints and footprints.

In addition to their humanlike hands and feet, raccoons also share a behavioral trait with us. The mothers care for the young for an extended period of time. Litters of two to five young are born once a year, in the spring, and the young raccoons remain with the mother throughout their first summer and winter. She teaches them where and how to forage for food and protects them from predators. Normally solitary, raccoons found in pairs or groups are usually a mother and her kits.

Raccoons are not only the most common carriers of rabies, but they are also the most frequently encountered rabid animal, in part because they live close to humans. In North Carolina, rabid raccoons outnumber all other rabid animals, such as skunks, foxes, bats, cats, and dogs, combined. Stay away from raccoons you see in broad daylight or are otherwise acting strangely. If you are bitten by a raccoon, you should be treated by a physician immediately because of the possibility of rabies transmission. Rabies and other diseases may, in fact, be what limits the population of raccoons, for they have few predators other than humans with guns and cars.

There are two Appalachian species of skunk. The striped skunk (*Mephitis mephitis*) is about the size of a house-cat, ranges from three to eleven pounds, and is black with a white stripe running from the nose to the tip of the tail. The width of the white stripe, which usually splits into two stripes along the back, varies among individuals, and some skunks have stripes so wide that they appear completely white (albinos also occur). Although widely distributed across North America, striped skunks tend to be more common in the mountains than along the coastal plain of the eastern states. The eastern spotted skunk (*Spilogale putorius*) is smaller and restricted to the Southern Appalachians, where it can be more numerous than the striped skunk. It, too, is mostly black but has an irregular pattern of white spots and stripes. It weighs no more than four pounds.

Skunks' contrasting markings advertise their unpalatability to most predators. When threatened, skunks are able to discharge musk from their anal glands. Their spray can accurately reach its target up to eighteen feet away. They do not, however, spray indiscriminately because the spray is expensive to produce. When threatened, they first stamp their feet, elevate their tail, click their teeth, growl, and hiss. Before they spray, they bend their body so that both head and anus face the intruder and then evert the openings to the glands. Striped skunks keep all their feet on the ground and bend sideways, but spotted skunks stand on their front feet and bend their anus over their head. It is true that a skunk is unable to spray if lifted by the tail before the glands are exposed, but being sure of the glands' position requires a closer examination than might be prudent. Some predators, however, have a poor sense of smell, undermining the protection that the musk affords. Great horned owls are the most common predators of skunks and frequently smell like their prey. Rehabilitators who rescue injured great horned owls sometimes have to segregate them from other animals until the smell wears off. Bathing a great horned owl with tomato juice, the most common antidote to skunk musk, is hard to imagine.

Although skunks are omnivores, they prefer insects, and up to 70 percent of their diet is animal matter. They are one of the few animals that will dig up an underground nest of yellow jackets in order to eat the larvae. They are beneficial in that their food consists mainly of animals that most people don't want around: mice, snakes, beetle grubs, and insects such as yellow jackets, but they may also eat birds and their eggs or any other small animal that they encounter. Berries form the principal plant portion of their diet. They are also more common than most people realize and live closer than we imagine. They normally live in hollow trees, rocky dens, or abandoned burrows, but they are comfortable under houses or barns. They are almost never active during daylight, preferring the hours of true darkness, and they remain active year-round except during very cold weather. The spotted skunk prefers moonless nights, perhaps because great horned owls are more active on moonlit nights when they can see their prey.

The Virginia opossum (*Didelphis virginiana*) is the only marsupial mammal in North America, more akin to the kangaroo than to the mammals in our region. They are abundant and, as is the case with raccoons, are frequently killed along roadsides as they scavenge for food at night. Opossums are omnivores, but they eat mainly animal matter such as roadkill, insects, snails, and salamanders. They are eaten by many carnivorous mammals, including humans. My great-grandfather, for instance, relished them as food, but only after they had been kept in a pen and fed on bread and milk for several days.

Opossums are about the size of a house cat and weigh around five pounds. They are generally gray in color, although young animals are frequently much darker. Their long, slender mouth contains fifty teeth, more than any other mammal in North America. They have a small braincase and a high sagittal crest (the bony ridge atop the skull where jaw muscles attach), which make the skull easy to identify.

Like other marsupials, the females have a pouch in which the young develop. As many as twenty offspring are born only twelve days after conception, but some die before they can crawl to the pouch. In the pouch, the tiny pups attach to one of twelve nipples. Two litters are produced each year, so the many opossums killed by cars and predators can be quickly replaced. Unlike placental mammals, female opossums have two uteri that unite in a single external opening. Males have a forked penis in order to deposit sperm in both sides of the reproductive tract. The forked penis gave rise to the myth that opossums impregnate the female through her nose.

In addition to using their numerous, sharp teeth, opossums can avoid becoming food for carnivores by "playing possum." When opossums are threatened, they first try to climb a tree to escape, using their prehensile tail and opposable thumbs on the hind feet. Although good climbers, they are slow walkers, with a maximum speed of four miles per hour (gray squirrels can move at a rate of seventeen miles per hour). They growl, hiss, and show their teeth but rarely bite. When all else fails to fend off a predator, they feign death by curling up on one side, half-closing their eyes, lolling their tongue from a partially open mouth, and secreting a foul-smelling fluid from their anal glands. All this is to convince predators, which normally rely on movement to elicit a hunting response, that the animal in front of them is carrion, not live prey.

Opossums are underappreciated when it comes to intelligence. In bright light, including car headlights, they panic and freeze, but that's because, as nocturnal animals, their sensitive eyes are blinded temporarily. When, however, intelligence tests are conducted in dim light, they outperform dogs, ranking close to the rather intelligent pigs!

## Rarely Seen Large Carnivores:
## Foxes, Coyotes, Bobcats, Mountain Lions, and Otters

Two foxes, one native and one introduced, occur in the Appalachians. The gray fox (*Urocyon cinereoargenteus*) is the native species. Although its back, legs, and tail are gray, its throat, belly, and the undersurface of its tail can be quite reddish. It can be confused with the red fox but can be positively identified by its black tail tip. Gray foxes prefer forested habitats, but they adapt readily to suburban areas where forests are broken by small grassy openings such as lawns. Gray foxes eat birds and small mammals like mice and rabbits, but they also eat berries and other fruits when they are available in summer and fall. They are good at climbing trees to reach these foods, and their shortened faces and muzzles make them look catlike.

The red fox (*Vulpes vulpes*) was introduced into the southeastern United States from Europe for sport hunting, but it may be native to the boreal forests of Canada. It is mostly reddish in color but has gray around its legs and the underside of its tail. The tail tip is white. Red foxes prefer more open habitats than gray foxes and are more often found in agricultural fields. Like gray foxes, they eat mice, rabbits, and other small mammals, as well as birds; while they eat some plant material, they rely less on it than do gray foxes, and they do not climb trees. Although they are an introduced species in the Southeast, they do not appear to threaten gray fox populations. Like some of our European weeds, they have naturalized into the landscape without significantly displacing other species.

Coyotes (*Canis latrans*) have become common in eastern North America over the last decade. Originally an animal of the western plains, coyotes have moved eastward for several reasons. For one, its competitors, the red wolf (*C. rufus*) and the gray wolf (*C. lupus*) have been extirpated from the region. (An attempt to reintroduce the red wolf into the Great Smoky Mountains National Park from 1991 to 1998 failed, and the remaining wolves were moved back into breeding facilities.) Second, as competition has decreased, food resources for coyotes have increased. They eat nearly anything they find, from plant material to roadkill. Since they rarely attack healthy adult livestock, their notoriety as livestock killers most likely is a result of their scavenging carcasses of dead animals and their attraction to newborn animals, which they can kill. They will also kill and eat small dogs and cats, which are in the normal size range of their prey. Third, their preferred habitat of semi-open land has grown: highways have opened grassy corridors through forested land, brushy borders of large agricultural fields and timbered forests have increased, and suburban developments have opened woodlands. No wonder coyotes have become as common in the East as in the West. There is an occasional rumor that wildlife officials intentionally released coyotes in the East, but the only coyotes that were intentionally released were introduced by hunters for sport (along with red foxes). Coyotes are perfectly capable of dispersing on their own!

Coyotes are tan in color and can easily be confused with domestic dogs or foxes. The best way to distinguish the three is by their behavior: when coyotes run or walk, they hold their tails down; dogs hold theirs straight up; and foxes hold theirs horizontally. Coyotes can weigh anywhere between twenty and forty pounds. They're generally larger than foxes but smaller than a German shepherd. They are most vocal in fall, when the young of the year are learning to howl. The young are born in the spring and stay with their parents until the next spring. Occasionally they remain to help with the young born the following year but more often disperse before the next litter is born. Coyotes can interbreed with dogs and produce hybrid (nonfertile) offspring.

Bobcats (*Lynx rufus*) are the only common native cat in the eastern states. Rabbits and other small mammals are the predominant prey of bobcats, but small deer are also occasionally killed. In contrast to domestic cats, bobcats generally don't hunt birds. (One reason that most native birds are so easily killed by domestic cats is that they have so little experience with cats' hunting techniques.) Bobcats have short (bobbed) tails, which distinguishes them from mountain lions. They can be surprisingly long-legged and large, measuring as much as three feet long and weighing as much as twenty-five pounds.

The cougar (*Puma concolor*), also called mountain lion, catamount, puma, panther or painter, and tyger, has been extirpated from the east. (The endangered subspecies of the Florida panther is still with us, however). Unconfirmed sightings, especially from the Southern Appalachians and from southeastern swamps, suggest that a remnant population of cougars might be present, but no firm evidence has been provided. A few cougars have actually been collected east of the Mississippi, but they were escaped animals that had been kept as pets. Surprisingly, there is a brisk business in cougars and other large, potentially dangerous cats as "pets."

In 2007, the U.S. Fish and Wildlife Service began a scientific review of the evidence supporting the status of the eastern cougar (*P. concolor couguar*) as an endangered species. As part of the review, scientists evaluated the likelihood that a breeding population of eastern cougars exists, the possibility that western cougars are expanding their range eastward, and whether all confirmed sightings were of previously captive animals.

Captive animals may also partly explain the persistent rumor of black panther sightings in the Southern Appalachians. Several species of large cats, such as jaguars, leopards, and jaguarundi, sometimes produce melanistic, or dark, offspring. Jaguars and jaguarundi occur in South America and wander occasionally into Texas, but not beyond. Leopards occur only in Asia and Africa. A melanistic cougar, however, has never been bred or collected, and most breeders and biologists agree that black cougars do not exist. Thus, the black cats people have claimed to have seen cannot be cougars. Rather, they are likely escaped jaguars, leopards, or jaguarundi cats (from the pet trade). People may also be seeing a melanistic bobcat, for bobcats are frequently misidentified as cougars, and black bobcats have been collected in Florida, suggesting that they could occur elsewhere in the Southeast.

Cougars are huge cats. Their bodies can be up to six feet long and weigh as much as two hundred pounds. Their tail, which can be as long as three feet, distinguishes them from bobcats. Their primary prey is white-tailed deer. It seems that such large animals could not avoid detection, but cougars are notoriously shy and wary. Maybe there really are some big cats in these hills!

North American river otters (*Lontra canadensis*) were also extirpated from the Appalachian region. They were reintroduced into the Great Smoky Mountains National Park in 1986 and to the French Broad and other western North Carolina rivers in 1992. About one hundred total animals were released in western North Carolina, and quite successfully, for sightings of river otters are now frequent in the region. Otters are always associated with water, and their primary food is fish. The presence of otters is often indicated by their piles of scat, composed almost exclusively of fish scales. Their webbed back feet and sleek body make them agile swimmers. They are dark brown, can reach four feet in length, and weigh up to twenty-three pounds. River otters have few natural enemies but were seriously affected by human activities. Like the related American mink, they were trapped for their beautiful and dense fur, poisoned by water pollution, and excluded by environmental damage to river and riparian habitat. Now they are protected.

## Eyeshine of Nocturnal Animals

Many mammals have adapted to nocturnal activity by enhancing their vision. Their highest period of activity tends to be centered at dusk and dawn, when there is some light in the sky. This pattern is called crepuscular and is more common among nocturnal animals than activity in the dead of night.

Most nocturnal mammals see chiefly shades of gray. They are unable to see colors because most of the light-sensitive cells in the retinas of their eyes are rods instead of cones. Cones can detect color (see sidebar "Color Perception in Humans" on page 125), but rods cannot. Rods are activated by lower levels of light than are cones, allowing animals whose eyes are composed primarily of rod cells to see better in dim light. Color is less important to them than the contrast afforded by a black-and-white image.

Most nocturnal mammals have a reflective layer of cells, called a tapetum lucidum, behind the retina of their eyes. The tapetum reflects the pattern of light that passes through the retina back onto it, superimposing the reflected pattern precisely onto the incoming pattern, thus stimulating more rod cells and improving the image. When a car's headlights or a flashlight shines into such eyes, the reflection produces eyeshine. Different animals have different color eyeshines, depending on the cell arrangement of the tapetum. Deer, for example, have silvery white eyeshine, rabbits have red eyeshine, and opossums have pinkish eyeshine. Most cats and dogs have green eyeshine, but some breeds of dogs may have yellow or even blue eyeshine.

Birds, reptiles, and some mammals lack a tapetum lucidum but also produce eyeshine when their eyes are illuminated. Their eyeshine is red because their heavily vascularized retinas reflect the red color of the eyes' myriad capillaries. Alligators have red eyeshine, which hunters use to locate and identify them at night, owls also have red eyeshine, and human eyeshine shows up as "red-eye" in flash photographs. The eyes of arthropods such as insects, spiders and shrimp have individual facets rather than smooth retinas but can still produce eyeshine. Their white eyeshine, like the red eyeshine of alligators, can be used to locate them at night.

## Rabies Epidemics

The incidence of rabies can change dramatically within a few years. In North Carolina in the 1990s there was a dramatic increase in the number of animals identified with the disease, from only 10 animals in 1990 to a peak of 879 in 1997 and a continued high level through 2006. Three different epidemics, one centered in Florida, one in West Virginia, and one farther north, merged to create a large-scale epidemic in eastern North America that stretched from Florida to Nova Scotia. Another large-scale epidemic occurred during the 1950s and 1960s, when the majority of animals identified were domestic dogs and cats. As a result of that epidemic, rabies vaccinations for dogs and cats are now required.

Rabies is a virus that attacks the brain and nervous system of an animal and is invariably fatal. It is transmitted by a bite from or contact with the nervous tissue of an infected animal. Carnivores and scavengers are most often affected because they eat moribund or dead animals. In the current epidemic, the animal most commonly affected is the raccoon. Other carnivores, such as bats (which eat mosquitoes carrying mammalian blood), foxes, skunks, house cats, bobcats, and domestic dogs are also frequent carriers, probably infected when they kill and eat slow-moving or sick animals that are themselves infected with the disease. Herbivores, such as cows, horses, goats, groundhogs, deer, beaver, and rabbits are less common carriers, but any mammal can carry the disease. Opossums, while numerous, frequently encountered, and scavengers, very rarely transmit rabies (only one case has been recorded in North Carolina). As marsupial mammals, they have a low body temperature, making it difficult for the virus to prosper. Rabies has not been confirmed in squirrels, shrews, or mice.

The frequency with which humans come in contact with a mammal almost certainly influences the data. It is likely, for instance, that any cow, horse, or goat that died from rabies would be recorded. On the other hand, since humans don't generally encounter dead mice in the wild, the incidence of rabies among mice would not be recorded. Similarly, the incidence of rabies in coyotes, who are carnivores and thus expected to be frequent carriers, in the recent rabies epidemic is rare (only five cases reported in North Carolina). This is probably related both to their normally secretive nature and to their relatively recent arrival in the eastern states. Since the 1990s, as an attempt to prevent the epidemic from spilling out of the Appalachians into the Midwest, edible packets of oral vaccine have been dropped all along the Appalachian chain from eastern Ohio to Alabama.

## Cycles of Bird Irruptions

In some years, birds of the far north suddenly appear in large numbers farther south than expected during the winter season. These birds are referred to as irruptive species because they irrupt into more southern regions (and erupt, or exit, more northern ones). Finches and grosbeaks frequently irrupt.

Their movement south appears to be related to a failure of wild food sources in the north. There is no relation between the coldness of the winter and the movement south. Some years we see a few finches, some years we see almost none, but some years we see droves.

Finches are adapted to eating seeds. While other birds have to hammer a seed with their beak to open it, finches can hold the seed and peel it while it is held in their beak. They sit still, only their beaks and tongues moving as they peel seed after seed. They remind me of a virtuoso banjoist who creates an incredible progression of notes while barely moving anything more than his fingers. Less efficient seed-eaters such as chickadees spend a lot of energy in picking up a seed with their beak, transferring it to their feet to hold it against a tree branch, or wedging it into a crevice, then hammering away at it. Sometimes the seed flies out of their grasp, so they have to fly toward the ground after it.

If your feeder has been overrun with finches because of the winter's finch irruption, how can you tell the species apart? You'll definitely need a field guide. "Finching" is a bit of a challenge, for winter males of different species look similar, and the brownish females are even harder to tell apart.

Purple finches (*Carpodacus pupureus*) are often confused with house finches (*C. mexicanus*). Both are about the size of tufted titmice, which are regulars at most feeders. Male purple finches look as if they have been dipped in the juice of pokeberries or raspberries. Their rump patch, the point at which the tail joins the body of the bird, is particularly colorful, as is the head, but the whole bird is splashed with color. Male house finches are also a sort of purple, but the color is redder and more restricted to the head and rump patch. The females of both species are finely streaked in gray, but the female purple finch has thicker, darker stripes on the face.

Their habitats are also different. House finches are native to the western United States and Mexico, were introduced into New York City in 1940, and have spread up and down the eastern seaboard. They have become permanent residents in urban and suburban areas. Purple finches require more natural, forested areas, nesting in Canada and the Northern Appalachians and wintering farther south. The introduced house finches are often afflicted by a bacterial disease (mycoplasmal conjunctivitis) that causes their eyes to swell shut with crusty sores. It first appeared in the winter of 1993–94. Affected finches die because they are unable to see and cannot find food, avoid predators, or fly without crashing into objects.

Eastern goldfinches (*Carduelis tristis*) and pine siskins (*C. pinus*) can also be confused with each other. Both are smaller than purple finches, close in length to chickadees, but slimmer. In the spring, male goldfinches are unmistakable with bright yellow, black, and white feathers, but in the winter their colors are more subdued. Unlike pine siskins, goldfinches are not streaked but smoothly colored in light yellow and gray. Pine siskins have brown streaks on their heads, backs, and breasts. The key to identifying them is their yellow wing bar, which is easiest to see when the bird opens its wings in an aggressive display toward other birds. Goldfinches are present year-round, but pine siskins are only present during the winters of finch irruptions. The pine siskins are the smallest finch but are the most aggressive birds at the feeder.

While these four birds are the most common finches, we are occasionally treated to winter flocks of evening grosbeaks (*Coccothraustes vespertinus*) that come down from the north. They are large, colorful birds the size of a cardinal and have large, heavy beaks similar to the cardinal's. The bold pattern of yellow, black, and white on their feathers makes them hard to miss or to mistake for another bird. Both the males and females are similarly patterned, although the female is somewhat duller in color. On a few occasions when I was as a child growing up in South Carolina, evening grosbeaks would arrive, and I would dash to the windows to watch them and imagine what other fantastic birds must live "up north."

## Birdsong and the Cycles of Territoriality versus Foraging Flocks

Birdsong is a phenomenon of the natural world that many people find engaging and beautiful. The melodious warbles of wrens, the cheery notes of titmice, and the friendly drawls of cardinals all sound like music to our ears. To birds, however, these pretty songs are more akin to the blaring bugles and throbbing drums of martial music. They are announcements of strength and declarations of resources. Birds advertise their territories and attract females while warning other males not to trespass with their calls. And it is the male who does most of the singing.

In the spring, birds are very vocal, for that is when most are establishing territories. There is often a brief period of quiet when the young have fledged (left the nest) and the adults are too busy feeding them to squabble with other birds and attract attention to their young. For birds that breed a second time in the season, calls may increase again, but in fall and winter, birds generally speak to one another in more muted tones.

By listening to the cycle of birdsong, as it moves from loud and boisterous to quiet and placid, we can also track birds' flocking and territoriality, the cyclic seasonal patterns of bird behavior. In spring, pairs of birds actively defend their territories. For instance, I've noticed a pair of northern cardinals aggressively chasing any other cardinal, male or female, from the feeder outside my window during spring months. By summertime, however, the adults have their young in tow and tolerate the presence of other cardinals. In fall and winter, several males and females will occur together in feeding flocks without any sign of aggression. Similarly, mixed flocks of chickadees, titmice, nuthatches, and woodpeckers move through the trees together and suddenly descend on the feeder en masse. In spring, however, tufted titmice, who are among the most aggressive of birds, will vigorously defend the feeder from others. In winter, hundreds of blackbirds, starlings, grackles, cowbirds, and crows will also gather in groups. Wintertime is the season of flocking; spring is the season of territoriality.

Birds defend territories because they need resources to raise young. Males usually establish the initial limits of the territory, then attract mates based on females' perceived quality of resources within the territory. In the winter, however, these territories break down; many birds migrate away and others move around locally. Because the birds are more exposed to predators, with less cover from deciduous leaves and annual plants, and because it may be advantageous to work together to find less common food sources, birds forage in flocks. Perhaps the flocks also initiate pair-bonding; joining the flock is a way for birds to "meet" other members of the species in the local area.

In the Southern Appalachians, one of the most common wintertime foraging flocks is composed of titmice, chickadees, nuthatches, and woodpeckers. Tufted titmice (*Baeolophus bicolor*) are common year-round residents in most of eastern North America, where they are regulars at bird feeders. Their body is mostly a soft blue-gray, and their underside is white and highlighted by two pale peach patches just under the wings. Their jaunty crest is erect most of the time, and their large black eyes scrutinize their surroundings carefully. Titmice are quite vociferous. Invariably in late winter and early spring, someone will ask me to identify the bird that awakens him every morning with a relentless, boisterous "PEter-PEter-PEter!" During nesting season, titmice also make a constant, high-pitched, "see-see-see" call that can get under your skin if you let it.

Chickadees often travel with tufted titmice. Southern chickadees are almost all Carolina chickadees (*Poecile carolinensis*), but at high elevations and farther north, the black-capped chickadee (*P. atricapillus*), a slightly larger species, occurs. Both species have a black cap and a black bib. Their back and wings are gray, their chest is creamy white, and the sides of their head are a cleaner color of white; black-caps also have a patch of white in their gray wings. Chickadees make a two-note call, described as "feee-bee," but their classic "chick-a-dee-dee-dee" is more common; the songs of the black-caps differ slightly from those of the Carolinas.

White-breasted nuthatches (*Sitta carolinensis*) are another common resident in eastern North America. They remind me of wind-up toys as they descend headfirst down tree trunks, taking a few steps to the right, then turning to make a few to the left as they search for insects hidden in the crevices of bark. They also readily come to feeders for seeds, which they wedge into tree-bark crevices and hammer open. Both males and females have a gray back and a white face and belly, but the male's cap is darker. Their nasal "yank-yank-yank" calls enhance their cute, toylike appearance. Brown creepers (*Certhia americana*) also work tree trunks for insects, but, unlike nuthatches, they spiral *up* the tree, from base to crown. They are brown on back and white underneath and, while present in the Appalachians year-round, are most visible in winter.

Titmice, chickadees, and nuthatches are cavity nesters. Chickadees and white-breasted nuthatches often excavate their own nests, but tufted titmice do not. Downy woodpeckers (*Picoides pubescens*), in addition to frequently traveling in these winter flocks, often unintentionally provide homes for their compatriots, for woodpeckers usually excavate a new nest hole each season. Titmice, chickadees, and nuthatches also readily accept constructed boxes. Because tufted titmice love to line their nests with hair, they will go to great extremes to get it, even plucking it from resting mammals such as dogs or humans! These perky birds are hard to miss whether you're hiking or at home.

Another common winter flock of mixed species is composed of those foragers more often found on the ground, birds of the sparrow family. They include northern cardinals, dark-eyed juncos, and eastern towhees. Carolina wrens also frequently join these ground-feeding birds even though wrens are not seed-eaters. They probably appreciate the protection of the other members of the flock. All scratch noisily among the leaves as they search for seeds or insects.

Imagine for a moment that you are on an exotic vacation in a tropical locale. Just ahead of you, you hear a friendly, slurred "cheer, cheer, cheer!" from a thicket of tangled plants and vines. You search for the songster, and suddenly he pops into view: a brilliant red bird, feathers the scarlet of a priest's robes. A jaunty crest gives him an added look of elegance, and the black quadrangle surrounding his orange-red bill makes that seed-cracking device stand out like Pinocchio's nose. Bird-watchers are delighted by a glimpse of similar shockingly colored birds such as scarlet ibis, roseate spoonbills, vermilion flycatchers, and painted buntings, but this spectacular bird rarely excites such wonder because he is so common; we call him "*The* redbird." We seldom appreciate beauty we see every day, but long instead for the unusual. Next time you see a male northern cardinal (*Cardinalis cardinalis*), take a moment to enjoy him for what he is, a fantastically colored bird.

Whether highlighted against the dazzling winter snow or perched in the dark green of a summer rhododendron, that incredible red always attracts my attention when northern cardinals arrive. They are usually the first birds to the feeder in the morning and the last to leave at night, lingering until the shadows are long and the sun has left the sky. As darkness falls, red is the first color of the spectrum to disappear. In semidarkness, red turns black, thus providing a perfect camouflage. (Incidentally, that is why many deep-sea fish and invertebrates are red; in the twilight depths, red looks black.)

Cardinals have large bills that make them adept at cracking seeds, so they readily accept sunflower seed placed in feeders. They are, in fact, often among the first birds to visit a new feeder. In the summertime, when raising young, they feed almost exclusively on insects such as caterpillars and beetles. In wintertime, however, seeds form the bulk of their diet. Cardinals are common in wooded and suburban areas throughout the eastern United States, rarely making it as far north as Canada. Because they prefer low, brushy growth, they prosper near fencerows, scrubby edges of lots, or piles of brush. They often build their nests close to the ground in such tangles.

Cardinals are family-oriented, and a mated pair remains together year-round. When breeding season begins, the male often plies the female with tidbits of food, sometimes even peeling the seeds he offers her. Not only does this behavior serve to unite the pair, but it demonstrates that the male has located a high-quality territory. If you have a feeder, watch for this behavior, for it signals that nesting will soon follow. Once the young have fledged, the male cares for that brood while the female starts to lay eggs for the next clutch. They raise two or three broods each season.

Eastern towhees (*Pipilo erythrophthalmus*) occur year-round throughout most of the Southeast, but they are summer breeders in the northern states. Males are elegant, with a black head, back, and tail, a white belly, and rufous sides. Females are similar, but duller, with brown replacing black. Both have brilliant red eyes, from which their scientific name is derived ("erythro" means red). They feed and scratch on the ground, in the manner of other sparrows, but they flick open their tail to expose its white corners. Their call is a loud "tow-hee" or, as southerners describe it, "jor-ree!" My grandfather, a South Carolinian, called the birds jorees, and I have since learned that the Cherokees also called them jorees. In fact, they referred to the Blue Ridge Mountains as the Joree Mountains.

Dark-eyed juncos (*Junco hyemalis*) and white-throated sparrows (*Zonotrichia albicollis*) are winter birds in most of the South, nesting farther north. My grandfather called them snow-hoppers. In the Southern Appalachians, we usually see them in their spring plumage and hear their breeding songs before they depart, and a few juncos even nest at high elevations. I once surprised a junco on its nest at Devil's Courthouse and collected a window-killed junco in juvenal plumage, both in Transylvania County, North Carolina.

Eastern phoebes (*Sayornis phoebe*) are not seed-eaters but, like wrens, eat insects. Phoebes are flycatchers. To catch insects, they wait on an exposed perch, zip out to grab an insect in midair, then return to the perch, where they sit with tail bobbing up and down. This characteristic bobbing tail, coupled with the fact that they are nearly tame and will land near people, provides a good opportunity to identify them. Phoebes are rather plain in appearance, grayish in color with a dark head. They nest in buildings, barns, or under bridges and are year-round residents of the Southern Appalachians. The phoebe's nest is a cup of mud and moss, lined with animal hair.

Four different species of wren occur in the Appalachians, three commonly. The Carolina wren (*Thryothorus ludovicianus*) is the most conspicuous because it frequents woodlands and farm fields throughout the eastern United States, is abundant year-round, and is such a noisy, inquisitive little bird. Its white eyebrow and chunky brown body can only be confused with those of its rare cousin, Bewick's wren, which differs from the Carolina in its song and the white edges of its tail.

House wrens (*Troglodytes aedon*) occur across most of northern North America during spring and summer, which is their breeding season. In the winter, they migrate to the southeastern coast and into Mexico. They are slim, grayish brown birds, lacking the strong eye stripe of the Carolina. Males may start several nests, but the female chooses and finishes the final one, which may be another example of a female choosing a mate based on territorial resources. They have a bad reputation for puncturing the eggs of other birds, but this behavior is a natural form of population control, because it is more prevalent when there is increased competition for limited nesting sites. Few individual wrens actually do the puncturing. Their bold-ness and beautiful, bubbling song make them welcome in any case.

Both Carolina and house wrens are cavity nesters and, like eastern blue-birds, build their nests inside tree trunks or bird boxes. Unlike bluebirds, however, wrens are not picky, and often choose such unlikely places as coat pockets, open buckets, or even narrow spaces behind paint cans on a garage shelf. They do not mind being close to humans, even though this trait makes them more susceptible to being killed by pet cats.

Winter wrens (*Troglodytes troglodytes*), which winter throughout the South-east, may be found in the Appalachians year-round. This tiny, brown bird can easily be mistaken for a mouse as it hops along stream banks and heavily wooded ravines, rarely flying more than a few feet at a time. Unlike its cous-ins, the winter wren does not frequent human habitations. In the spring, if you're out in the forest near dawn or dusk, you just might hear a winter wren practicing his sustained, melodious, and effervescent song. It sounds as fluid and lively and gentle as the mountain brook he perches beside to sing.

Wrens have some of the most beautiful of songs. They are bold and bub-bly, punctuated with loud phrases and tempered by soft refrains. Early one morning, in a tree near my bedroom window, a tufted titmouse boldly pro-claimed his "PEter, PEter!" Next, an eastern phoebe struggled to recall the exact rendition of its own song. Instead of the clearly whistled "fee-bee," his call was slurred and off-key, more suited to a Saturday night than Sunday morning. Perhaps he had been eating soft and moldy berries, which had been transformed by the shine of the moon. When the Carolina wren began to sing, however, the chorus hushed as the principal stepped onto the stage.

Along with his ardent "tea-kettle, tea-kettle, tea-kettle" serenade, the exu-berant male Carolina wren filled the silence between each aria with a friendly, buzzing recitative clearly meant for a nearby female. I peeked from my win-dow to see his paramour perched on the porch railing, her head cocked to-ward the hidden soloist. She was listening, enrapt by what she heard. The *primo uomo* was not in view, but frantic scratchings and an increased volume suggested that he was trapped somewhere. Carefully easing open the win-dow, I extended my head. From the open, deep pocket of one of the back-packs hanging under the eaves, the virtuoso burst forth, his lady friend close behind as they flew off together to a nearby brush pile. Perched there upon the highest branch, buff chest expanded and tail alertly upright, the male's indignant songs chastised me for disturbing his courtship. With luck, such songsters will provide a season of magnificent performances for you, too.

## Reproductive Cycles that Begin in Winter:
## Wood Frogs and Great Horned Owls

Winter may seem like a strange time for an animal to reproduce, but a few of them get a head start by mating and laying eggs soon after the days begin to lengthen. Amphibians are not generally considered winter-hardy animals because most hibernate during the winter months, but one species, the wood frog (*Rana sylvatica*), awakens early to mate during the depths of winter. "Sylvan" means wooded or forested. "Rana" is the common group of large frogs.

Wood frogs gather in shallow ponds in late January or early February, usually just after a period of rain, to mate and lay their eggs. The adults are evident for a up to a week and are most conspicuous when the males call to advertise both their pond and their availability. The males' calls sound just like the quacking of ducks, and they call whenever it is warm enough for them to be active, which is usually beginning at midday. If you hear but don't see ducks on a pond in midwinter, look instead for frogs.

The males in the pond congregate around each female, pushing and jostling each other to be the first male to reach her. The dark male clasps the pinkish female around the belly and hangs on as other males try to dislodge him. When females are scarce, dozens of males may clasp onto one female and even onto each other, forming a mass of writhing frogs. As the female releases eggs, the males release sperm, and the cluster of males around each female ensures that all her eggs are fertilized, often by several different males.

Females lay eggs in fish-free ponds or even puddles. The eggs laid by one female are stuck together, and each egg looks like a tiny grape with a single black seed in the center, which is the developing embryo. This egg mass can be as big as a large grapefruit and is often attached to a piece of aquatic vegetation or a submerged stick. The frogs choose a shallow area that receives some warmth from the sun, which encourages the eggs to develop more rapidly. They are also attracted to other egg masses and tend to lay in clusters. If a late bout of cold temperatures causes the pond to ice over, the uppermost eggs freeze and are killed, but the eggs that remain deeper in the water do not freeze and usually survive.

The adult frogs are a different story. They can survive being frozen as long as their temperature does not drop below about 25° F. As they chill, they move dissolved sugar from their liver into their blood and tissues, which acts as antifreeze. During the cold winter months, the adult frogs hibernate by burrowing into the soil, where they are protected from extremes of temperature. Because of this ability to withstand cold temperatures, wood frogs are widely distributed in the northern states, in Canada, and all along the Appalachian Mountains.

The tadpoles develop rapidly when the water warms up, and the newly metamorphosed wood frogs often leave their natal pond shortly after other frogs begin to lay their eggs. Their early breeding period may be a way to minimize competition from other species of frog that utilize the ponds later in the season or to avoid the warmer-season predators such as snakes, turtles, and other frogs. The numerous tadpoles tend to cluster together near the surface of the water, where it is the warmest, and they drop down to the bottom shortly before they grow legs and metamorphose.

Adult wood frogs are easy to identify by sight as well as sound. They are tan with dark masks across their eyes (see Figure 1-2). Unlike many of our other frogs, they travel far from water when they are not breeding. In the forest, I have most often encountered the tiny froglets that have recently metamorphosed from tadpoles.

Great horned owls (*Bubo virginianus*) also mate and lay eggs in winter. Perhaps they do so because it is easier to find and catch their prey before trees leaf out, or maybe it just takes these large owls a long time to raise their young. Because skill at catching prey is learned, it may be advantageous for young predators to come of age when their prey is most abundant; young great horned owls learn to hunt in the spring and summer months and stay with their parents until fall.

Great horned owls are most vocal during the winter as they establish their territory, warn intruding owls of their claim, and court one another. Their voice has earned them the nickname "hoot owl." Their deep, resonant hoot is made up of (usually) five hoots in the pattern of "HOO, hoo-HOO, HOO, HOO." These hoots are calm and low but carry well, like a bass baritone's voice. Barred owls, another common "hoot owl," hoot in a pattern of eight hoots, and are more often heard in the summer. Their tenor voices are emphatic and forceful, calling what sounds like "Who cooks for you? Who cooks for y'all?" Both types of owls make other eerie sounds, especially near midnight, when what you imagine often trumps reality.

One summer evening several years ago, I was seated on the porch of our cabin in the woods, home again after a month or more away. Just as dusk descended and the outlines of tree trunks were blurring into ethereal forms, a huge, black, winged creature appeared immediately before me. Without a sound, it veered upward, wings outspread like an angel's or demon's, and then was gone. I nearly fell out of my chair, wondering if I'd had heart attack

FIGURE 5-7
The great horned owl (*Bubo virginianus*) has ear tufts and yellow eyes. Its huge talons and large size allow it to hunt prey such as house cats and skunks.

and was being visited by the angel of death, before logic overcame my initial, primeval response. It was a great horned owl that had gotten used to perching under the porch roof while we were gone, startled when it came to use its accustomed perch and found it already occupied!

Part of the eeriness of owls is their absolute silence while flying. This quietness is achieved by special soft feathers, even covering the feet, that muffle the sound of their wing beats. This silent approach is especially important for catching nimble prey. I frequently find their beautiful feathers in the field around our garden. The smallest feathers are pretty mixtures of brown, white, and tan, while the larger feathers have a dirty white background with bold dark brown bars across them. All the feathers have soft edges with wispy downlike bases.

Great horned owls occur throughout the Appalachians, and they prefer edge locations where field and forest come together. They occur in habitats similar to those of red-tailed hawks. They are sometimes spotted hunting from tree perches, where they look like big-headed, neckless hawks. The "horns" are tufts of feathers that stick up above their heads (Figure 5-7). In addition to the tufts, they have yellow eyes. As large as two feet tall with wingspans of up to five feet, they eat mostly whatever they want, from mammals and birds to reptiles and amphibians. Skunks are a favorite food. Because of their absolute fearlessness when attacking prey, they are also called "tiger owls." They have been documented eating everything from hawks to house cats. Incidentally, crows mob owls, hawks, and snakes because these animals are predators on them. If you should hear a mob of crows calling, it is worth following their fuss, for it will often lead you to a great horned owl you might have otherwise missed.

## Climatic Conditions in Winter

The winter solstice, which marks the shortest day and the longest night of the year, occurs on December 20th or 21st. The winter season has the shortest days of the year, and December has the shortest average day length of any month, a little less than 10 hours. The average high temperature is 51°F, and the average low is 28°F. The average total precipitation is 4.5 inches over nine days, and 0.7 inches of snow fall.

On average, January is the coldest month of the year in the Great Smoky Mountains. January's average high is only 48°F, and the average low is 25°F. It is common in the Southern Appalachians to experience a severe cold spell during January in which the daily temperature does not reach above freezing for several days. Ponds freeze over, trickling waterfalls become ice monuments, and water squeezes up from the soil like fingers of some buried ice monster. January is also the snowiest month of the year. On average, 4.9 inches of total precipitation fall over nine days, and 3.6 inches of snow fall. Like the hurricanes of summer and droughts of fall, these occasional but damaging winter storms shape the forest for years to come. The average day length in January is 10 hours.

The month of February is still cold, but lengthening days — 11 hours of daylight on average — encourage plants and animals to prepare for the arrival of spring. The average high temperature is 52°F, and the average low is 26°F. On average, 4.3 inches of precipitation fall over eight days, with 1.4 inches of snow.

# Appendix

## Federal Public Lands in the Southern Appalachians

Blue Ridge Parkway: 469 miles, from Oconaluftee Ranger Station in Great Smoky Mountains National Park to Rockfish Gap in Shenandoah National Park

Skyline Drive: 105 miles, contained within Shenandoah National Park

Great Smoky Mountains National Park in North Carolina and Tennessee: 521,086 acres

Shenandoah National Park in Virginia: 197,439 acres

Jefferson–George Washington National Forest in Virginia: 690,100 acres

Monongahela National Forest in West Virginia: 935,000 acres

Cherokee National Forest in Tennessee: 640,000 acres

Pisgah National Forest in North Carolina: 510,119 acres

Nantahala National Forest in North Carolina: 531,303 acres

Chattahoochee National Forest in Georgia: 750,502 acres

Andrew Pickens Ranger District of the Sumter National Forest in South Carolina: 84,000 acres

Talladega/Shoal Creek Ranger District of the Talladega National Forest in Alabama: 233,585 acres.

# References

The references listed here are not meant to be an exhaustive list of all publications relating to the biology and natural history of the Southern Appalachian region. Rather, these references were used to provide the factual details included in this book. The references are divided into regional field guides, general references, and scientific journal articles for those who want to dig a little deeper.

## REGIONAL FIELD GUIDES

Bentley, S. L. *Native Orchids of the Southern Appalachian Mountains*. Chapel Hill: University of North Carolina Press, 2000.

Covell, C. V. *A Field Guide to the Moths of Eastern North America*. Boston: Houghton Mifflin Co., 1984.

Duncan, W. H., and M. B. Duncan. *Trees of the Southeastern United States*. Athens: University of Georgia Press, 1988.

Hemmerly, T. E. *Appalachian Wildflowers*. Athens: University of Georgia Press, 2000.

Kricher, J. C. *A Field Guide to Eastern Forests*. Peterson Field Guide Series. Boston: Houghton Mifflin Co., 1988.

Levi, H. W., and L. R. Levi. *A Guide to Spiders*. New York: Golden Press, 1968.

Linzey, D. W. *A Natural History Guide to Great Smoky Mountains National Park*. Knoxville: University of Tennessee Press, 2008.

Martof, B. S., W. M. Palmer, J. R. Bailey, J. R. Harrison III, and J. Dermid. *Amphibians and Reptiles of the Carolinas and Virginia*. Chapel Hill: University of North Carolina Press, 1980.

Radford, A. E., H. E. Ahles, and C. R. Bell. *Manual of the Vascular Flora of the Carolinas*. Chapel Hill: University of North Carolina Press, 1968.

Rohde, F. C., R. G. Arndt, D. G. Lindquist, and J. F. Parnell. *Freshwater Fishes of the Carolinas, Virginia, Maryland, and Delaware*. Chapel Hill: University of North Carolina Press, 1994.

Simpson, M. B. *Birds of the Blue Ridge Mountains: A Guide for the Blue Ridge Parkway, Great Smoky Mountains, Shenandoah National Park, and Neighboring Areas*. Chapel Hill: University of North Carolina Press, 1992.

Smith, R. M. *Wildflowers of the Southern Mountains*. Knoxville: University of Tennessee Press, 1998.

Stewart, K. G., and M. R. Roberson. *Exploring the Geology of the Carolinas: A Field Guide to Favorite Places from Chimney Rock to Charleston*. Chapel Hill: University of North Carolina Press, 2007.

Turner, N. J., and A. F. Szczawinski. *Common Poisonous Plants and Mushrooms of North America*. Portland: Timber Press, 1991.

Wagner, D. L. *Caterpillars of Eastern North America*. Princeton: Princeton University Press, 2005.

Weakley, A. S. *Flora of the Carolinas, Virginia, Georgia, Northern Florida, and Surrounding Areas*. Chapel Hill: UNC Herbarium, North Carolina Botanical Garden, University of North Carolina, 2008. <www.herbarium.unc.edu>.

Webster, W. D., J. F. Parnell, and W. C. Biggs Jr. *Mammals of the Carolinas, Virginia, and Maryland*. Chapel Hill: University of North Carolina Press, 1985.

## GENERAL REFERENCES

Bartram, W. *Travels of William Bartram: Naturalist's Edition*. Edited by F. Harper. Athens: University of Georgia Press, 1998.

Bent, A. C. *Life Histories of North American Birds of Prey*. Vol. 1. New York: Dover Publications, 1961.

Bir, R. E. *Growing and Propagating Showy Native Woody Plants*. Chapel Hill: University of North Carolina Press, 1992.

Boligino, C. *The Appalachian Forest: A Search for Roots and Renewal*. Mechanicsburg: Stackpole Books, 1998.

Boligino, C., and J. Roberts. *The Eastern Cougar: Historic Accounts, Scientific Investigations, and New Evidence*. Mechanicsburg: Stackpole Books, 2005.

Bonta, M. *Appalachian Summer*. Pittsburgh: University of Pittsburgh Press, 1999.

Brooks, M. *The Appalachians*. Boston: Houghton Mifflin Co., 1965.

Campbell, N. A., and J. B. Reece. *Biology*. 7th ed. San Francisco: Pearson-Benjamin Cummings, 2005.

Camuto, C. *Another Country: Journeying Toward the Cherokee Mountains*. (About the reintroduction of red wolves.) New York: Henry Holt and Co., 1997.

Constantz, G. *Hollows, Peepers, and Highlanders: An Appalachian Mountain Ecology*. 2nd ed. Morgantown: West Virginia University Press, 2004.

Davis, D. E. *Where There Are Mountains: An Environmental History of the Southern Appalachians*. Athens: University of Georgia Press, 2000.

Eisner, T. *For Love of Insects*. Cambridge: Harvard University Press, 2003.

Ellison, G., and E. Ellison. *Blue Ridge Nature Journal: Reflections on the Appalachian Mountains in Essays and Art*. Charleston: Natural History Press, 2006.

Findlay, W. P. K. *Fungi: Folklore, Fiction, and Fact*. Surrey: The Richmond Publishing Co., 1982.

Hallowell, B. G. *Mountain Year: A Southern Appalachian Nature Notebook*. Winston-Salem: Blair Publishing, 1998.

Howell, P. K. *Medicinal Plants of the Southern Appalachians*. Mountain City: Botanologos Books, 2006.

Kaufman, K. *Lives of North American Birds*. Boston: Houghton Mifflin Co., 1996.

Nolt, J. *A Land Imperiled: The Declining Health of the Southern Appalachian Bioregion.* Knoxville: University of Tennessee Press, 2005.

Phillips, H. R. *Growing and Propagating Wildflowers*. Chapel Hill: University of North Carolina Press, 1985.

Proctor, M., P. Yeo, and A. Lack. *The Natural History of Pollination*. Portland: Timber Press, 1996.

Ruppert, E. E., R. S. Fox, and R. D. Barnes. *Invertebrate Zoology: A Functional Evolutionary Approach*. 7th ed. Belmont: Thomson-Brooks/Cole, 2004.

Schaechter, E. *In the Company of Mushrooms: A Biologist's Tale*. Cambridge: Harvard University Press, 1998.

Teale, E. W. *The Strange Lives of Familiar Insects*. New York: Dodd, Mead, & Co., 1962.

Thoreau, H. D. *Autumn*. New York: Houghton Mifflin Co., 1892.

Van Gelderen M., P. C. De Jong, H. J. Oterdoom, and T. R. Dudley. *Maples of the World*. Portland: Timber Press, 1994.

Weidensaul, S. *Mountains of the Heart*. Golden, Colo.: Fulcrum, 1994.


SCIENTIFIC JOURNAL ARTICLES AND TEXTS

Adams, D. A., and J. S. Hammond. "Changes in Forest Vegetation, Bird, and Small Mammal Populations at Mount Mitchell, North Carolina: 1959/62 and 1985." *Journal of the Elisha Mitchell Society* 107 (1991): 3–12.

Amman, G. D., and C. F. Speers. "Balsam Woolly Aphid in the Southern Appalachians." *Journal of Forestry* 63 (1965): 18–20.

Asner, G. P., J. M. O. Scurlock, and J. A. Hicke. "Global Synthesis of Leaf Area Index Observations: Implications for Ecological and Remote Sensing Studies." *Global Ecology & Biogeography* 12 (2003): 191–205.

Beattie, A. J. *The Evolutionary Ecology of Ant-Plant Mutualism*. Cambridge: Cambridge University Press, 1985.

Beattie, A. J., and N. Lyons. "Seed Dispersal in Viola (Violaceae): Adaptations and Strategies." *American Journal of Botany* 62 (1975): 714–22.

Brower, L. P. "Canary in the Cornfield: the Monarch and the Bt Corn Controversy." *Orion Magazine*, 2001.

———. "Forest Thinning Increases Monarch Butterfly Mortality by Altering the Microclimate of the Overwintering Sites in Mexico." In *Decline and Conservation of Butterflies in Japan II*, edited by S. A. Ae, T. Hirowatari, M. Ishii, and L. P., 33–44. Brower Proceedings International Symposium on Butterfly Conservation. Osaka, Japan, 1994.

Brower, L. P., D. R. Kust, E. Rendon-Salinas, E. G. Serrano, K. R. Kust, J. Miller, C. Fernandez del Rey, and K. Pape. "Catastrophic Winter Storm Mortality of Monarch Butterflies in Mexico in January 2002." In *Monarch Butterfly Biology and Conservation*, edited by K. M. Oberhauser and M. Solensky, 151–66. Ithaca: Cornell University Press, 2004.

Brower, L. P., G. Castilleja, A. Peralta, J. Lopez-Garcia, L. Bojorquez-Tapia, S. Diaz,

D. Melgarejo, and M. Missrie. "Quantitative Changes in Forest Quality in a Principal Overwintering Area of the Monarch Butterfly in Mexico: 1971 to 1999." *Conservation Biology* 16 (2002): 346–59.

Chittka, L., and T. F. Döring. "Are Autumn Foliage Colors Red Signals to Aphids?" *PLOS Biology* 5 (2007): 1640–44.

Cipollini, M. L., and D. J. Levey. "Why are Some Fruits Toxic? Glycoalkaloids in *Solanum* and Fruit Choice by Vertebrates." *Ecology* 78 (1997): 782–98.

Conner, E. F., and M. P. Taverner. "The Evolution and Adaptive Significance of the Leaf-Mining Habit." *Oikos* 79 (1997): 6–25.

Covington, M. F., and S. L. Harmer. "The Circadian Clock Regulates Auxin Signaling and Responses in *Arabidopsis*." *PLOS Biology* 5 (2007): 1773–84.

Cramer, J. M., M. L. Cloud, N. C. Muchhala, A. E. Ware, B. H. Smith, and G. B. Williamson. "A Test of the Bicolored Fruit Display Hypothesis: Berry Removal with Artificial Fruit Flags." *Journal of the Torrey Botanical Society* 130 (2003): 30–33.

Dunning, D. C., L. Acharya, C. B. Merriman, and L. D. Ferro. "Interactions Between Bats and Arctiid Moths." *Canadian Journal of Zoology* 70 (1992): 2218–23.

Eisner, T., K. Hicks, M. Eisner, and D. S. Robson. "'Wolf-in-Sheep's Clothing' Strategy of a Predaceous Insect Larva." *Science* 199 (1978): 790–94.

Eisner, T., and S. Nowicki. "Spider Web Protection Through Visual Advertisement: Role of the 'Stabilimentum.'" *Science* 219 (1983): 185–87.

Enright, J. T. "Sleep Movements of Leaves: In Defense of Darwin's Interpretation." *Oecologia* 54 (1982): 253–59.

Farner, D. S. "Annual Rhythms." *American Review of Physiology* 47 (1985): 65–82.

Gaddy, L. L. "A New *Hexastylis* (Aristolochiaceae) from Transylvania County, North Carolina." *Brittonia* 38 (1986): 82–85.

Galloway, J. N., F. J. Dentener, D. G. Capone, E. W. Boyer, R. W. Howarth, S. P. Seitzinger, G. P. Asner, C. C. Cleveland, P. A. Green, E. A. Holland, D. M. Karl, A. F. Michaels, J. H. Porter, A. R. Townsend, and C. J. Vörösmarty. "Nitrogen Cycles: Past, Present, and Future." *Biogeochemistry* 70 (2004): 153–226.

Gargiullo, M. B., and E. W. Stiles. "Chemical and Nutritional Differences Between Two Bird-Dispersed Fruits: *Ilex opaca* and *Ilex verticillata*." *Journal of Chemical Ecology* 17 (1991): 1091–1106.

Garshelis, D. L., and M. R. Pelton. "Activity of Black Bears in the Great Smoky Mountains National Park." *Journal of Mammalogy* 61 (1980): 8–19.

Gerhardt, H. C., and R. Huber. *Acoustic Communication in Insects and Anurans: Common Problems and Diverse Solutions*. Chicago: University of Chicago Press, 2002.

Givnish, T. J. "Adaptive Significance of Evergreen vs. Deciduous Leaves: Solving the Triple Paradox." *Silva Fennica* 36 (2002): 703–43.

Graveland, J., R. van der Wal, J. H. van Balen, and A. J. van Noordwijk. "Poor Reproduction in Forest Passerines from Decline of Snail Abundance on Acidified Soils." *Nature* 368 (1995): 446–48.

Handel, S. N., and A. J. Beattie. "Seed Dispersal by Ants." *Scientific American* 263 (1990): 58–64.

Heinrich, B. "Bee Flowers: A Hypothesis on Flower Variety and Blooming Times." *Evolution* 29 (1975): 325–34.

Hiers, J. K., and J. P. Evans. "Effects of Anthracnose on Dogwood Mortality and Forest Composition of the Cumberland Plateau (USA)." *Conservation Biology* 11 (1997): 1430–35.

Herberstein, M. E., C. L. Craig, J. A. Coddington, and M. A. Elgar. "The Functional Significance of Silk Decorations of Orb-Web Spiders: A Critical Review of the Empirical Evidence." *Biological Review* 75 (2000): 649–69.

Horton, C. C. "A Defensive Function for the Stabilimenta of Two Orb Weaving Spiders (Araneae, Araneidae)." *Psyche* 87 (1980): 13–20.

Hristov, N. I., and W. E. Conner. "Sound Strategy: Acoustic Aposematism in the Bat-Tiger Moth Arms Race." *Naturwissenschaften* 92 (2005): 164–69.

Hurlbert, A. H., S. A. Hosoi, E. J. Temeles, and P. W. Ewald. "Mobility of *Impatiens capensis* Flowers: Effect on Pollen Deposition and Hummingbirds Feeding." *Oecologia* 105 (1996): 243–46.

James, R. L. "Some Hummingbird Flowers East of the Mississippi." *Castanea* 13 (1948): 97–109.

Janzen, D. H. "Why Fruits Rot, Seeds Mold, and Meat Spoils." *The American Naturalist* 3 (1977): 691–713.

Jetz, W., D. S. Wilcove, and A. P. Dobson. "Projected Impacts of Climate and Land-Use Change on the Global Diversity of Birds." *PLOS Biology* 5 (2007): 1211–19.

Johnson, R. A., M. F. Willson, J. N. Thompson, and R. I. Bertin. "Nutritional Values of Wild Fruits and Consumption by Migrant Frugivorous Birds." *Ecology* 66 (1985): 819–27.

Krebs, C. J. "Population Cycles Revisited." *Journal of Mammalogy* 77 (1996): 8–24.

Kunz, T. H., and M. B. Fenton, eds. *Bat Ecology*. Chicago: University of Chicago Press, 2003.

Lapointe, L. "How Phenology Influences Physiology in Deciduous Forest Spring Ephemerals." *Physiologia Plantarum* 113 (2001): 151–57.

Lasiewski, R. C. "Oxygen Consumption of Torpid, Resting, Active, and Flying Hummingbirds." *Physiological Zoology* 36 (1963): 122–40.

Leake, J. R. "Plants Parasitic on Fungi: Unearthing the Fungi in Myco-Heterotrophs and Debunking the 'Saprophytic' Plant Myth." *Mycologist* 19 (2005): 113–20.

Lepczyk, C. A., K. G. Murray, K. Winnett-Murray, P. Bartell, E. Geyer, and T. Work. "Seasonal Fruit Preferences for Lipids and Sugars by American Robins." *The Auk* 117 (2000): 709–17.

Levi, M. A. B., I. S. Scarminio, R. J. Poppi, and M. G. Trevisan. "Three-Way Chemometric Method Study and UV-Vis Absorbance for the Study of Simultaneous Degradation of Anthocyanins in Flowers of the *Hibiscus rosea-sinensys* Species." *Talanta* 62 (2004): 299–305.

Losey, J. E., L. S. Rayor, and M. E. Carter. "Transgenic Pollen Harms Monarch Larvae." *Nature* 399 (1999): 214.

Lyman, C. P., J. S. Willis, A. Malan, and L. C. H. Wang. *Hibernation and Torpor in Mammals and Birds*. New York: Academic Press, 1982.

Macior, L. W. "The Pollination Ecology of *Dicentra cucullaria.*" *American Journal of Botany* 57 (1970): 6–11.

Meiners, S. J., and E. W. Stiles. "Selective Predation on the Seeds of Woody Plants." *Journal of the Torrey Botanical Society* 124 (1997): 67–70.

Mitchell, R. G., G. D. Amman, and W. E. Waters. "Balsam Woolly Aphid." *Forest Pest Leaflet 118.* USDA Forest Service, 1970.

Nilsen, E. T. "Why Do Rhododendron Leaves Curl?" *Arnoldia: The Magazine of the Arnold Arboretum* 50 (1990): 30–35.

Nishida, R. "Sequestration of Defensive Substances from Plants by Lepidoptera." *Annual Review of Entomology* 47 (2002): 57–92.

Oak, S. W. "From the Bronx to Birmingham: Impact of Chestnut Blight and Management Practices on Forest Health Risks in the Southern Appalachian Mountains." *Journal of the American Chestnut Foundation* 16 (2002): 32–41.

Oberrath, R., and K. Böhning-Gaese. "Phenological Adaptation of Ant-Dispersed Plants to Seasonal Variation in Ant Activity." *Ecology* 83 (2002): 1412–20.

Ostfeld, R. S., C. G. Jones, and J. O. Wolff. "Of Mice and Mast." *Bioscience* 46 (1996): 323–30.

Otero, J. T., and N. S. Flanagan. "Orchid Diversity—Beyond Deception." *Trends in Ecology and Evolution* 21 (2006): 64–65.

Paoletti, C., and K. E. Holsinger. "Spatial Patterns of Polygenic Variation in *Impatiens capensis,* a Species with an Environmentally Controlled Mating System." *Journal of Evolutionary Biology* 12 (1999): 689–96.

Pollard, K. M. "Population Growth and Distribution in Appalachia: New Realities." In *Demographic and Socioeconomic Change in Appalachia.* Washington, D.C.: Population Reference Bureau, 2005.

Potter, D. A., and T. W. Kimmerer. "Do Holly Leaf Spines Really Deter Herbivory?" *Oecologia* 75 (1988): 216–21.

———. "Inhibition of Herbivory on Young Holly Leaves: Evidence for the Defensive Role of Saponins." *Oecologia* 78 (1989): 322–29.

Powell, J. A. "Interrelationships of Yuccas and Yucca Moths." *Trends in Ecology and Evolution* 7 (1992): 10–15.

Pudlo, R. J., A. J. Beattie, and D. C. Culver. "Population Consequences of Changes in an Ant-Seed Mutualism in *Sanguinaria canadensis.*" *Oecologia* 146 (1980): 32–37.

Real, L., ed. *Pollination Biology.* New York: Academic Press, 1983.

Risser, P., and G. Cottam. "Influence of Temperature on the Dormancy of Some Spring Ephemerals." *Ecology* 48 (1967): 500–503.

Robinson, R. A., J. A. Learmonth, A. M. Hutson, C. D. Macleod, T. H. Sparks, D. I. Leech, G. J. Pierce, M. M. Rehfisch, and H. Q. P. Crick. "Climate Change and Migratory Species." *British Trust for Ornithology (BTO) Research Report* 414, 2005.

Root, T. L., J. T. Price, K. R. Hall, S. H. Schneider, C. Rosenzweig, and J. A. Pounds. "Fingerprints of Global Warming on Animals and Plants." *Nature* 421 (2003): 57–60.

Rossell, I. M., and J. M. Kesgen. "The Distribution and Fruiting of Red and Black

Chokeberry (*Aronia arbutifolia* and *A. melanocarpa*) in a Southern Appalachian Fen." *Journal of the Torrey Botanical Society* 130 (2003): 202–5.

Rust, R. W. "Pollination in *Impatiens capensis* and *Impatiens pallida* (Balsaminaceae)." *Bulletin of the Torrey Botanical Club* 104 (1977): 361–67.

Schemske, D. W., M. F. Willson, M. M. Melampy, L. J. Miller, L. Verner, K. M. Schemske, and L. B. Best. "Flowering Ecology of Some Spring Woodland Herbs." *Ecology* 59 (1978): 351–66.

Schnurr, J. L., R. S. Ostfeld, and C. D. Canham. "Direct and Indirect Effects of Masting on Rodent Populations and Tree Seed Survival." *Oikos* 96 (2002): 402–10.

Schupp, E. W. "Quantity, Quality and the Effectiveness of Seed Dispersal by Animals." *Plant Ecology* (1993) 107–8.

Sebeok, T. A., ed. *How Animals Communicate*. Bloomington: Indiana University Press, 1977.

Shelley, R. M. "Annotated Checklist of the Millipedes of North Carolina (Arthropoda: Diplopoda), with Remarks on the Genus *Sigmoria* Chamberlin (Polydesmida: Xystodesmidae)." *Journal of the Elisha Mitchell Society* 116 (2000): 177–205.

———. "Centipedes and Millipedes with Emphasis on North American Fauna." *The Kansas School Naturalist* 45 (1999): 3–15.

Sherratt, T. N., D. M. Wilkinson, and R. S. Bain. "Why Fruits Rot, Seeds Mold, and Meat Spoils: a Reappraisal." *Ecological Modelling* 192 (2006): 618–26.

Simmons, R. B., and S. E. Weller. "What Kind of Signals do Mimetic Tiger Moths Send? A Phylogenetic Test of Wasp Mimicry Systems (Lepidoptera: Arctiidae: Euchromiini)." *Proceedings of the Royal Society of London B* 269 (2002): 983–90.

Simpson, M. B., Jr. "Ecological Factors Contributing to the Decline of Bewick's Wren as a Breeding Species in the Southern Blue Ridge Mountain Province." *Chat* 42 (1978): 25–28.

Snow, B., and D. Snow. *Birds and Berries: A study of an ecological interaction*. Calton, England: T&AD Poyser, 1988.

Soule, J. D. "Bewick's Wren." *The Nature Conservancy Species Management Abstract*. 1992.

Sparks, T. H., and A. Menzel. "Observed Changes in Seasons: An Overview." *International Journal of Climatology* 22 (2002): 1715–25.

Stiles, E. W. "Fruit Flags: Two Hypotheses." *The American Naturalist* 120 (1982): 500–509.

———. "Patterns of Fruit Presentation and Seed Dispersal in Bird-Disseminated Woody Plants in the Eastern Deciduous Forest." *The American Naturalist* 116 (1980): 670–88.

Storey, K. B., and J. M. Storey. "Frozen and Alive." *Scientific American* (1990): 92–97.

Thien, L. B. "Mosquito Pollination of *Habenaria obtusata* (Orchidaceae)." *American Journal of Botany* 56 (1969): 232–37.

Thien, L. B., and F. Utech. "The Mode of Pollination in *Habenaria obtusata* (Orchidaceae)." *American Journal of Botany* 57 (1970): 1031–35.

Thompson, J. N., and M. F. Willson. "Disturbance and the Dispersal of Fleshy Fruits." *Science* 200 (1978): 1161–63.

———. "Evolution of Temperate Fruit/Bird Interactions: Phenological Strategies." *Evolution* 33 (1979): 973–82.

Tsahar, E., J. Friedman, and I. Izhaki. "Impact on Fruit Removal and Seed Predation of a Secondary Metabolite, Emodin, in *Rhamnus alaternus* Fruit Pulp." *Oikos* 99 (2002): 290.

Visser, M. E., C. Both, and M. M. Lambrechts. "Global Climate Change Leads to Mistimed Avian Reproduction." *Advances in Ecological Research* 35 (2004): 89–110.

Vornanen, M. "Basic Functional Properties of the Cardiac Muscle of the Common Shrew (*Sorex araneus*) and Some Other Small Mammals." *Journal of Experimental Biology* 145 (1989): 339–51.

Wang, L. C. H., and M. W. Wolowyk. "Torpor in Mammals and Birds." *Canadian Journal of Zoology* 66 (1988): 133–37.

Warren, W. S., and V. M. Cassone. "The Pineal Gland: Photoreception and Coupling of Behavioral, Metabolic, and Cardiovascular Circadian Outputs." *Journal of Biological Rhythms* 10 (1995): 64–79.

Weller, S. J., N. L. Jacobsen, and W. E. Conner. "The Evolution of Chemical Defenses and Mating Systems in Tiger Moths (Lepidoptera: Arctiidae)." *Biological Journal of the Linnean Society* 68 (1999): 557–78.

Whigham, D. F., and M. McWethy. "Studies on the Pollination Ecology of *Tipularia discolor* (Orchidaceae)." *American Journal of Botany* 67 (1980): 550–55.

White, P. F., E. Buckner, J. D. Pittillo, and C. V. Cogbill. "High-Elevation Forests: Spruce-Fir Forests, Northern Hardwoods Forests, and Associated Communities." In *Biodiversity of the Southeastern United States: Upland Terrestrial Communities*, edited by W. H. Martin, S. G. Boyce, and A. C. Echternacht, 305–38. New York: John Wiley and Sons, 1993.

Witmer, M. C., and P. J. Van Soest. "Contrasting Digestive Strategies of Fruit-Eating Birds." *Functional Ecology* 12 (1998): 728–41.

Wolff, J. O. "Population Fluctuations of Mast-Eating Rodents are Correlated with Production of Acorns." *Journal of Mammalogy* 77 (1996): 850–56.

Yokogawa, T., W. Marin, J. Faraco, G. Pézeron, L. Appelbaum, J. Zhang, R. Rosa, P. Mourrain, and E. Mignot. "Characterization of Sleep in Zebrafish and Insomnia in Hypocretin Receptor Mutants." *PLOS Biology* 5 (2007): 2379–97.

Zahner, R., and S. M. Jones. "Resolving the Type Location for *Shortia galacifolia* T.&G." *Castanea* 48 (1983): 163–73.

# Index

Page numbers in bold refer to figures.

*Abies*, 5, 167
Acadian Orogeny, 31
Accessory pigments, 123–24
*Accipiter. See* Hawks
*Acer. See* Maples
*Achaearanea tepidariorum*, 160
Acid deposition, 168–69, 175
Acoustic warning, 98
*Actaea pachypoda*, 83, **Plate 23**
*Actias luna*, 98
Adelgids, 153, 167–69, **Plate 49**
*Agkistrodon contortrix*, 114, 115,
    **Plate 40**
*Agraulis vanillae*, 94
Albino squirrel, 137
*Albizia julibrissin*, 12, **13**
Alder, 43, 155–56
Alleghanian Orogeny, 32
Allen's hummingbird, 11
Alligator snapping turtle, 117
*Allium*, 44–46, 47
All Taxa Biodiversity Inventory, 7
*Alnus serrulata*, 43, 155–56
*Amelanchier arborea*, 36–37, **Plate 1**
American Holly. *See* Holly
*Amianthium muscitoxicum*, 46
Ammons, A. R., 7
Angiosperms, 165
*Anolis carolinensis*, **18**, 112
*Antheraea polyphemus*, 96–98, **97**
Anthocyanin, 124
Ants, 44, 48, 51–53, 76, 78, 88, 156, **157**
Aphids, **59**, 90, 127–28, 153, 155–57,
    167–69, **Plate 49**
*Apis mellifera. See* Honeybees

*Aplectrum hyemale*, 55, 172
Appalachian mountain formation, 4,
    10, 31–33
Apples, 130
Aquatic insects, 25–26
Arachnids, 157–60
*Araneae. See* Spiders
*Archilochus colubris*, 11, 41, 60, 66–68, 71,
    74, 138
Arctiidae moth family, 98
*Argiope aurantia*, 160
*Arisaema triphyllum*, 48, **Plate 6**
*Armillaria*, 122–23
*Asclepias*, 93, **129**, 149–50
Asheville, N.C., 2, 3, 147
Asian multicolored ladybug, 152–54, 168,
    **Plate 49**
Asteraceae, 151–52
*Auricularia*, 177
Autumn ladies' tresses. *See* Ladies'
    tresses orchids
Auxin, 19
Azalea, 41

*Baeolophus bicolor*, 141, 194, 196, 200
Ballooning by spiders, 158
Balsam fir, 5, 167
Balsam Grove, N.C., 104
Balsam woolly adelgid, 169
Banner Elk, N.C., 155
Barred owl, 143
*Bartonia*, 85, 126
Bats, 10, 11, 13, 20, 63, 97–98, 144–48,
    **146**, 182, 193
*Battus philenor*, 74, 94, 95, **Plate 15**

Beans. *See* Legumes

Bearded dragons, 106

Beechdrops, 59

Beech trees, 56, 59

Beehive, 59–61

Beetles: larva of, 86, 93, 187, **Plate 26**; predatory, 168

Beggar's ticks, 129

Biltmore House, 147

Biological clock, 10

Biological calendar, 10, 15–17

Bioluminescence, 105–7, 122–23

Bird feeders, 181, 185, 194

Bird's nest fungi, 86, **Plate 28**

Black bears, 20, 58, 89, 131, 180–82, 185

Blackberry, 130, 131

Black cherry, 63, 79

Black gum, 132

Black locust, 41, 43–44

Black panther, 191

Black rat snake, 4, 108, 116

Black trumpets, 86, 121

Black widow spider, 159

*Blarina brevicauda*, 183

Bloodroot, 51, **52, Plate 8**

Bluebead lily, 47

Blueberry, 130

Bluebird, 65–66, 134, 200

Blue butterfly, eastern tailed, 94

Blue ghost firefly, 104–5, **104**

Bluegill, 24

Blue grosbeak, 66

Blue Ridge Parkway, 1, **3**, 76, 126, 136, 142, 151, **Plate 50**

Blunt-leaved orchid, 83

Bogs, 3, 77–78, 117, 134

*Bombycilla cedrorum*, 135, 170

*Bombyx mori*. *See* Silkworm moth

Bobcats, 190–91

*Boraria stricta*, 107–8

Boxelder, 39

Box turtle, 117–18

Brevard, N.C., 132, 137, 143

Bridal bouquet, 173

British soldiers, 177

Broad-winged hawks, 141–42, 149

Brook trout, 24, 27, 46

Brown creepers, 197

Brown recluse spider, 159–60

*Bubo virginianus*. *See* Great horned owl

Bullbats, 139–40

Bumblebees, 41, 54, 60, 82, 88, 103, **Plate 9, Plate 35**

Burrowing: in snakes, 113; in moles/ shrews, 29–30, 183

*Buteo*. *See* Hawks

Butterflies, 10, 16, 41, 60, 71, 74, 93–96, **95**, 148–52, 156, 159, **Plate 15, Plate 33, Plate 45**

Butyl mercaptan, 14

Caesar's Head State Park, 32, 142

*Calopteryx maculata*, 102

Caloric value of fruits, 131–32

*Calostoma cinnabarina*, 86–87

Camouflage, 95–96, 98, 103, 108–9, 140, 156, 198, **Plate 35**

*Canis*, 189–90, 193

*Cantharellus*, 86, 121, 122

*Caprimulgus*, 140

Cardinal (*Cardinalis*), 66, 74, 141, 195, 196, 197–98

Cardinal flower, 7, 60, 66, 74–75, **Plate 15, Plate 47**

*Carduelis*, 193–95

Carnivorous plants, 76–78, 110, **Plate 18, Plate 19, Plate 20**

Carolina anole, **18**, 112

Carolina chickadee, 195–97

Carolina wren, 197, 199–200

Carotenoid, 124, 125

Carpenter bird, 109

*Carpodacus*, 193–95

*Carya illinoinensis*, 136

*Castanea*, 135–36

Cat, domestic, 63, 138, 190, 192, 200, 204

Catawba rhododendron, 78–79

Catbird, 135

Caterpillars, 61–65, 79–80, 93–96, **95**, 99, 148–50, **149**, 154–55, 156, 159, **Plate 33, Plate 46**

*Cathartes aura*, 142

*Catharus fuscescens*, 62, 134

Cedar waxwing, 135, 170

Centipede, 107

*Ceriodaphnia lacustris*, 23

*Certhia americana*, 197

Chanterelle, 86, 121, 122

*Charlotte's Web*, 160

Chattanooga, Tenn., 144

*Chelydra serpentina*, 117

Cherokees, 2, 75–76, 109, 128, 134, 138

Chestnut, 135–36

Chickadee, 195–97

*Chimaphila maculata*, 173

Chinquapin, 136

Chipmunk, 21, 131, 137

Chlorophyll, 58, 123–24; lack of, 59, 74, 80

*Chordeiles minor*, 139–40

Christmas fern, 174

Christmas trees, 167

*Chrysemys picta*, 117

*Chrysopa*, 156

Chuck-will's-widow, 140

Cicada, 28, **29**, 99, 102

Circadian rhythm, 10–12

*Cladina* and *Cladonia*, 177

Clinton's lily (*Clintonia*), 47

Clover, 41, 42, **43**

Clubmosses, 4, 32, 85, 173–75, **176**

Coal, 32, 175

*Coccothraustes vespertinus*, 195

*Coccyzus americanus*, 65

Cocklebur, 129

Collembola, 178, **179**

Colony collapse disorder, 61

Color perception, 125

Composite flowers, 151–52

*Condylura cristata*, 183

Cone cells, 125, 192

Conifers, 5, 32, 164–69

*Conopholis americana*, 57–59, **59**

Cooper's hawk, 141

Copperhead, 114, 115, **Plate 40**

Cordyceps, 86, **Plate 26**

*Cornus florida*, 36, 65, 130–33

Cougar, 190–91

Cowbird, brown-headed, 63

Coyote, 189–90, 193

*Cuscuta*, 73–74, **73**, **Plate 14**

Crab spider, 103, **Plate 35**

Craggy Gardens, 79

Cranefly orchid, 55, 56, 83, 172

*Craterellus*, 86, 121

Crawfish, 186

Crickets, 12, **15**, 99–101; as prey, 91

*Crotalus horridus*, 108, 114, **115**

Crow, 63, 141, 196, 204

*Cryphonectria parasitica*, 136

*Cryptobranchus alleganiensis*, 109–10

Crystal Falls, Mich., 123

Cyanide, 107, 108, 130

*Cyathus striatus*, 86, **Plate 28**

Cycles: daily, 9–12; seasonal, 9, 115–26; long-term, 10, 26–33

*Cypripedium*, 5, 54, 55, 85, **Plate 9**

Daddy-longlegs, 160

Damselfly, 102

*Danaus plexippus*. *See* Monarch butterflies

*Daphnia lumholtzi*, 23

Dark-eyed junco, 197, 199

Darwin, Charles, 12, 18–19, 76

Day length, 15–20

Deer. *See* White-tailed deer

Deer mouse. *See* Mice

*Dendrolycopodium obscurum*, 175, **176**

Descartes, 18

*Desmodus rotundus*, 146–47

Devil's Courthouse Overlook, 199

*Diadophis punctatus*, 116–17

*Didelphis virginiana*. *See* Opossum

*Diphasiastrum digitatum*, 175

*Discula destructiva*, 133

Dodder, 73–74, **73**, **Plate 14**

Doll's eyes, 83, **Plate 23**

Dogs: sensitivity to smell, 14

Dogwood anthracnose, 133

Dogwoods, 36, 65, 130–33

Downy woodpeckers, 197

Dowsing, 127

Dragonflies, 25, 101–2, 121, **Plate 34**

Drone fly, 90, **Plate 31**

*Drosera rotundifolia*, 76, **Plate 18**

Ducks, 201

*Dumetella carolinenesis*, 135
DuPont State Forest, N.C., 104
Dutchman's pipe, 94

Ebony jewelwing damselfly, 102
Echolocation, 13, 97–98, 144–48, 182–83
Elaiosome, 51–52, **52**
*Elaphe obsoleta*, 4, 108, 116
Elkmont, Tenn., 104
Elytra, 103, 104, 153
*Epargyreus clarus*, 94
*Epifagus virginiana*, 59
*Epigaea repens*, 80, 173
Equinox, 16, 69, 161
*Ericaceae. See* Heath family
*Eristalis tenax*, 90, **Plate 31**
*Erythronium*, 44, 46, 51, 72
Escarpment, Blue Ridge, 2, 32
*Euchaetias egle*, 150
*Eumenes fraternus*, 92
*Euonymus americanus*, 84, **Plate 24**
Evening grosbeak, 195
*Everes comyntas*, 94

Fabre, J. H., 64
*Fagus grandifolia*, 56, 59
*Falco sparverius*, 141
Fall webworms, 65, 79–80, 159
False hellebore, 46
False morels, 57
Femme fatale fireflies, 106
Fence lizard, 18, 111
*Feniseca tarquinius*, 156
Ferns, 4, 32, 173–75
Field garlic, 46
Finches, 193–95
Fire ants, 53
Fireflies, 15, 103–6
Firs, 5, 167
Fish, 6, 23–25, 27, 46, 101, 110, 191
Fishing spiders, 160
Flame azalea, 41
Flying squirrel, 5, 177, 183–84
Fly poison, 46
Foliar fruit flag, 132, 133
Forebrain, 10–11, 18
Fossil, 107, 178

Fox, 131, 183, 189
Foxfire, 122
*Frankia alni*, 43
Fraser fir, 167
Fraser magnolia. *See* Magnolia
French Broad heartleaf, 53
French Broad River, 7, 191
Fritillary butterfly, 94
Frogs, 7, 12, 14, 16, 68, 100, 109, 111, 117, 201–2, **Plate 12**
Frostbite, 164
Fruit dispersal, 38, 40, 44, 48, 51–53, 84, 128–38
Fungi, 56–57, **57**, 84–87, 121–23, 133, 136, 156, 177–78, **Plate 25**, **Plate 26**, **Plate 27**, **Plate 28**, **Plate 29**

Galax, 50, 172
Galls, 127–28
Gargoyles, 147–48
Garlic, 44–47
Garter snake, 108
*Gentianella quinquefolia*, 126, **Plate 43**
Gentians, 85, 121, 126, **Plate 43**
Gingko tree, 165
Ginseng, 75, **Plate 16**
Glaciation effects, 1, 5, 10
Glass lizards, 112
*Glaucomys*, 5, 177, 183–84
Global warming effects, 22, 53, 58, 169
Glowworm, 106
*Glyptemys*, 117
Goatsuckers, 140
Golden-crowned kinglets, 139
Goldenrod, 92, 103, 151
Goldenseal, 75
Goldfinch, 194
*Goodyera pubescens*, 56, 172
Grasses, 39, 47
Grasshopper, 100
Grass of Parnassus, 76
Gray, Asa, 50
Gray bats, 144
Gray catbird, 135
Gray fox, 189
Gray squirrel, 63, 108, 131, 132, 135, 137–38, 184, 185, 188, **Plate 37**

Gray wolf, 189

Great golden digger wasp, 91

Great horned owl, 7, 98, 143–44, 187, 202–4, **203**

Great Smoky Mountains National Park (GSMNP), 2, 3, 7, 17, 40, 69, 104, 112, 119, 161, 168–69, 189, 191

Grenville Orogeny, 31

Grosbeaks, 66, 195

Groundhog, 20, 179–80

Ground-pine, 175, **176**

Guano, 145

Gymnosperms, 165, 173

*Gyromitra*, 57

*Halesia tetraptera*, 7, 37, **Plate 2**

*Hamamelis virginiana*, 121, 126–28, **127**, 156

*Harmonia axyridis*, 152–54, 168, **Plate 49**

Harvester butterfly, 156

Harvestmen, 160

Hawks, 140–43, 149, 203

Heartleaf, 53, 94

Hearts-a-bustin', 84, **Plate 24**

Heath family, 78–80, 85, 173

Hellbender, 109–10

Hemlocks, 167–68, **Plate 50**

Hemlock woolly adelgids, 153, 167–68, **Plate 49**

*Hexastylis rhombiformis*, 53, 94

Hickory, 136

Hibernation, 20–21, 145–47, 179–82

Holly, 65, 133–35, 170–71

Honeybees, 41, 59–61, 71, 88, 90, **Plate 10**, **Plate 30**

Honeydew from aphids, 156

Honey mushrooms, 122–23

Hooded warbler, 62, **Plate 11**

Hoot owl. *See* Great horned owl

*Hormaphis hamamelidis*, **127**

Hormones, 10–11, 17–19

Hornets, bald-faced, 88–89

House finch, 194

House spider, 160

House wren, 199

Hover flies, 90, 156

Human colonization of Appalachians, 2

Hummingbird, 11, 41, 60, 66–68, 71, 74, 138

Humongous fungus, 123

*Huperzia lucidula*, 175

Hurricane effects, 161

Hyams, George, 50

*Hydrastis canadensis*, 75

*Hylocichla mustelina*, 5, 62, 134

*Hypercompe scribonia*, 155

*Hyphantria cunea*, 65, 79–80, 159

*Hypopitys monotropa*, 59, 80, **Plate 22**

Hypothalmus, 10–11, 18

Ice effects, 6, 163–66, 172, 204, **Plate 48**

Ichneumonids, 93

*Ilex. See* Holly

*Impatiens. See* Jewelweed

Inchworms, 96, 159

Indian-pipe, 59, 80

Indigo bunting, 65

Invasive species, 63, 153, 170

*Ipomoea*, 73

Iris, 52, **52**

Ironweed, 151

Irruptions of birds, 193–95

*Isotrema macrophyllum*, 94

Jack-in-the-pulpit, 48, **Plate 6**

Jack-o'-lantern mushrooms, 122

Jacobson's organ in snakes, 113

Jelly fungus, 177

Jewelweed, 49, 66, 71–74, **Plate 13**

Jocassee, 50

Joe-pye weed, 151

Joree, 197–99

Joyce Kilmer Memorial Forest, 40

*Juglans*, 135–37

*Junco*, 197, 199

Juneberry, 36–37, **Plate 1**

*Juniperus virginiana*, 135

*Kalmia latifolia*, 78–79, 161, 171, **Plate 21**

Katydid, 100

Keowee River, 50

Kestrel, 141

Kinglets, 139

Kudzu, 41, 42

Lacewings, larval, 156
Ladies' tresses orchids, 56, 126, **Plate 44**
Ladybugs, 152–54, 168, **Plate 49**
*Laetiporus sulphureus*, 86, **Plate 27**
Largemouth bass, 24
*Laricobius nigrinus*, 168
*Latrodectus*, 159
Leaf color change, 123–26
Legumes, 12, 41–44, 94
Leopard, 190–91
*Libellula lydia*, 101–2
Lichens, 176–78, 185
Light spectrum, 123
*Ligustrum sinense*, 170
Lilies, 47
*Limenitis*, 93–94, **95**
*Lindera benzoin*, 131–32
*Liriodendron tulipifera. See* Tulip trees
Little brown bat, 145, **146**
Lizards, 18, 111–13, **Plate 39**
*Lobelia cardinalis. See* Cardinal flower
Locust leaf miners, 44
*Lontra canadensis*, 7, 191
*Loxosceles reclusa*, 159–60
Luna moth, 98
*Lygaeus kalmii*, 150
*Lynx rufus*, 190–91

*Macrochelys temminckii*, 117
*Magicicada*, 28
Magnolia, 5, 36, 40–41, 121, 131–32
*Malacosoma americanum*, 63, 80, 159
Malheur National Forest, Oreg., 122
Mapleleaf arrowwood, 133
Maples, 4, 36, 37–39, **38**, 56, 125, 126,
  **Plate 42**
*Marmota monax*, 20, 179–80
Masting, 27–29, 111, 135
Mayfly, **25**
Medicinal plants, 45, 72, 75–76, 128, 173,
  175, **Plate 16**, **Plate 17**
*Megascops asio*, 143
Melatonin, 17–18
*Melittia cucurbitae*, 90, 91, **Plate 32**
*Mephitis mephitis. See* Skunk
Metabolism, 11, 12, 20–21, 144, 147, 148
Mice, 21, 98, 114, 123, 146, 183, **184**, 187

Michaux, Andre, 50
*Microtus pinetorum*, 29–31, **30**
Migration, 10, 20, 132, 134, 140–42,
  148–52
Milkweed bugs, 150
Milkweeds, 93, **129**, 149–50
Millipedes, 106–8, **107**, **Plate 36**
Mimicry, 89–91, 93–94, **95**–96, 98, 108,
  **Plate 33**
Mimosa, 12, **13**
*Misumena vatia*, 103, **Plate 35**
*Mitchella repens*, 173, **174**
Moccasin flower, 5, 54, 55, 85, **Plate 9**
Mockingbird, 135
Mole, 183
Monarch butterflies, 10, 16, 93–94, 148–
  52, **149**, **Plate 45**
Monocots, 47
*Monotropa/Monotropsis*, 80
Morels (*Morchella*), 56, **57**
Morning glory, 73
Mosquito, 77, 82–83, 101
Moths, 10, 14, 78, 81–82, **82**, 96–99, **97**,
  144, 150, 154–55, **Plate 32**
Mountain laurel, 78–79, 161, 171, **Plate 21**
Mountain lion, 190–91
Mount Mitchell, 2, 32, 168
Mount Pisgah Inn, 136
Mud-dauber, 92–93, **92**
Mushrooms, 56–57, **57**, 84–87, 121–23,
  177–78
Mussels, 6
*Mutinus elegans*, 85, 121
Mutualisms, 42, 43, 176–77
Mycelium, 56, 84, 87, 122
Mycoplasmal conjunctivitis, 194
Mycorrhizae, 56, 80, 84–85, 126
*Myotis. See* Bats

*Narceus americanus*, **107**
Native Americans. *See* Cherokees
*Nerodia sipedon*, 114–15, **Plate 41**
New River, 7
Newt, 110
Nightjars/nighthawks, 139–40
Nitrogen as fertilizer, 42–44, 80, 145
Nitrogen-fixing bacteria, 43

Nocturnal communication, 12–15; adaptations in bats, 144
Nodding ladies' tresses. *See* Ladies' tresses orchids
Norway spruce, 167
*Notophthalmus viridescens*, 110, **Plate 38**
Nursery web spiders, 160, **Plate 47**
*Nyssa sylvatica*, 132

Oaks, 4, 39, 58, 122, 135–36, 138, 165, **Plate 3**
*Obolaria virginica*, 85, 126
Oconee bells, 49–51, 172, **Plate 7**
*Odontota dorsalis*, 44
*Oecanthus fultoni*, 100–101
*Old Farmer's Almanac, The*, 154
*Omphalotus olearius*, 122
*Oncopeltus fasciatus*, 150
*Ophiocordyceps melolonthae*, 86, **Plate 26**
Opiliones, 160
Opossum, 15, 89, 131, 188–89, 192, 193
Orb-weaver spiders, 157–59
Orchids, 5, 36, 47, 54–56, 76, 83, 85, 172, **Plate 44**
Organ pipe mud-dauber wasp, 92, 92–93
Otter, 7, 191
Ovenbird, 62
Owls, 143–44. *See also* Great horned owl
*Oxydendrum arboreum*, 79, 121, 125
Ozone damage, 169

Painted turtle, 117
*Panax quinquefolius*, 75, **Plate 16**
Pangea, 32
Panther, 190–91
*Papilio. See* Swallowtails
*Paraprociphilus tessellatus*, 155–57
*Parascalops breweri*, 183
Parasitism, 56, 58–59, 73–74, 80, 84, 86, 122
*Parmotrema*, 177
Partridge-berry, 173, **174**
*Passerina*, 65, 66
Pecan, 136
Pennsylvanian period, 173, 175
*Peromyscus. See* Mice
Petersen, R. T., 99

*Phallus ravenelii*, 121, **Plate 25**
*Phausis reticulata*, 104, 104–5
Phengodidae, 106
Pheromones, 96, 105, 153
Phloem, 20
Phoebe, 89, 199, 200
Pholcid spiders, 160
*Photinus*, 103–04
Photoperiod, 19
*Photuris pennsylvanicus*, 103
*Picea*, 5, 167
*Picoides pubescens*, 197
Pileated woodpecker, 135
Pine, 36, 165, 168
Pine processionary caterpillar, 64
Pinesap, 59, 80, **Plate 22**
Pine siskin, 194
Pink lady's slipper orchid. *See* Moccasin flower
*Pinus*, 36, **165**, 168
Pipevine swallowtail, 74, 94, 95, **Plate 15**
*Pipilo erythrophthalmus*, 197–99
Pipistrelle bat, 145
*Pipistrellus subflavus*, 145
Pipsissewa, 173
Pisgah National Forest, 145
Pitcherplants, 77–78, 110, **Plate 19**, **Plate 20**
Pit vipers, 114–15, 123
Plankters, 21
*Platanthera ciliaris*, 55
*Platanthera obtusata*, 83
*Plestiodon*, 111, **Plate 39**
Plethodontidae, 109
*Pneumodesmus newmani*, 107
*Poecile*, 195–97
Poison ivy, 48, 49, 72, 133, 135
Poisonous snakes, 114–15
Poisonous spiders, 159–60
Poison sumac, 49, 134
Pollination, 35–41, 52–54, 60, 71–74, 79, 80–82, 83, 88, 127, 152
Pollution of air and water, 6, 26, 169, 175, 178
Polyphemus moth, 96–98, **97**

Poppy, 51
Potter wasp, 92
Possum. *See* Opossum
Pretty lips, 86–87
Prism, 123
Privet, 170
*Procyon lotor. See* Raccoons
*Prunus serotina*, 63, 79
*Pseudacris crucifer*, 68, 100, **Plate 12**
*Pseudocneorhinus bifasciatus*, 171
*Pterophylla camellifolia*, 100
Puddling behavior, 94
Puffballs-in-aspic, 86–87
*Puma concolor*, 190–91
Purple finch, 194
Puttyroot, 55, 172
*Pyrrharctia isabella*, 154–55, **Plate 46**

*Quercus. See* Oaks

Rabies, 146, 185, 186, 192–93
Raccoons, 63, 108, 118, 131, 138, 185, 186,
    193
Railroad worms, 106–7
Rainbow trout, 24, 27
Rain crow, 65
Rainfall, 5–6
Ramps, 44–46, 47
*Rana sylvatica. See* Wood frog
Rattlesnake, 108, 114, **115**
Rattlesnake plantain, 56, 172
Red cedar, 135
Red eft, 110, **Plate 38**
Red fox, 189
Red maple. *See* Maples
Red oak, 135–38. *See also* Oaks
Red-spotted purple butterfly, 94, 95
Red spruce, 5, 167
Red-tailed hawk, 141
Red wolf, 189
*Regulus*, 139
Reindeer moss, 177
*Rhizobium*, 42, 43
Rhododendron, 41, 78–79, 161, 171–72,
    **171**
*Rhus*, 133–34, **134**
*Rhyniognatha hirsti*, 178

*Rimelia*, 177
Ring-neck snake, **116**, 116–17
Ripening of fruits, 131
River otter, 7, 191
Roan Mountain, 79
Robber barons, 2
Robin, 134
*Robinia pseudoacacia*, 41, 43–44
Rock tripe, 177
Rod cells, 192
Rodinia, 31
Rosebay rhododendron. *See*
    Rhododendron
Royal jelly, 60
Ruby-crowned kinglets, 139
Ruby-throated hummingbird, 11, 41, 60,
    66–68, 71, 74, 138
Ruffed grouse, 108, 173
Ruffled lichens, 177
Running cedar, 175
Rush, 39

Salamanders, 3–5, 21, 109–11, 117, **Plate 38**
Salicylic acid, 94
Samara, 38
*Sanguinaria canadensis*, 51, **52**, **Plate 8**
*Sarracenia. See* Pitcherplants
Sarvis, 36–37
*Sasajicymnus tsugae*, 168
Sassafras, 121, 131
Saw-whet owl, 5
*Sayornis phoebe*, 89, 199, 200
*Scalopus aquaticus*, 183
*Sceloporus undulatus*, 18, 111
*Sciurus carolinensis. See* Gray squirrel
*Scorias spongiosa*, 156
Scorpion, 111
Screech owl, 143
Sedge, 39
*Seiurus aurocapillus*, 62
Sensitive-brier, 12
Serviceberry, 36–37, **Plate 1**
Shadbush, 36–37
Sharp-shinned hawk, 141
Shedding in snakes, 112
Shenandoah National Park, 3, 168
Shoemake, 133–34

*Shortia galacifolia*, 49–51, 172, **Plate 7**
Shrew, 11, 182–83
*Sialia sialis*, 65–66, 134, 200
*Sigmoria stenogon*, 107
Silk, 98, 99, 158–60
Silkworm moth, 96, 98, 99, 159
Silverbell, 7, 37, **Plate 2**
Silver-sided skipper, 94
*Sitta carolinensis*, 197
Skink, 111, **Plate 39**
Skunk, 14, 89, 186–87, 204
Slash pine, **165**
Sleep movements in plants, 12
Slime molds, 87, **Plate 29**
Snakes, 4, 14, 108, 111–17, **Plate 40,**
    **Plate 41**
Snapping hazelnut, 127
Snapping turtle, 117
Snow fleas, 178, **179**
Snowy tree cricket, 100–101
Social insects, 87–91
Sodium as a nutrient, 95
*Solenopsis invicta*, 53
Solstice, 16, 119, 204
Sooty mold, 156
*Sorex*, 11, 182–83
Sourwood, 79, 121, 125
Spanish moss, 177
Spermatophores in salamanders, 111
*Speyeria diana*, 94
*Sphex ichneumoneus*, 91
Spicebush, 131–32
Spiders, 91–93, 97, 157–60, **Plate 47**
Spider web: construction, 158–59; escape
    from, 97
*Spilogale putorius. See* Skunk
*Spilosoma virginica*, 154
*Spiranthes cernua. See* Ladies' tresses or-
    chids
Splash cups, 86, **Plate 28**
Spotted wintergreen, 173
Spring ephemerals, 44, 47, 52, 172
Spring peepers, 68, 100, **Plate 12**
Springtails, 178, **179**
Spruce, 5, 167
Spruce-fir forest, 139, 167, 168–69, 185
Squash borer moths, 90, 91, **Plate 32**

Squawroot, 57–59, **59**
Squirrel corn, 44
Squirrels. *See* Flying squirrel; Gray
    squirrel; Groundhog
Stapes, 112
Starlings, 66, 196
Stick-tights, 129
Stinkhorns, 85, 121, **Plate 25**
Stratification of ponds, 23
Strawberry bush, 84
*Strix varia*, 143
Sulfur shelf mushroom, 86,
    **Plate 27**
Sumac, 133–34, **134**
Sundew, 76, **Plate 18**
Swallow-tailed kite, 101
Swallowtails, 41, 60, 71, 74, 94, 95,
    **Plate 15, Plate 33**
Synchronous firefly, 103–4

Taconic Orogeny, 31
Tadpoles, 202
*Tamias striatus*, 21, 131, 137
Tannins, 136, 138
Tapetum lucidum, 192
*Tegeticula yuccasella*, 81–82, **82**
Tent caterpillars, 63, 80, 159
*Terrapene carolina*, 117–18
Territoriality in birds, 195–96
*Thalassaphorura encarpata*, 179
Thermometer, 171
Thoreau, Henry David, 33, 128
Thrush, 5, 134, 138, 141
*Thryothorus*, 197, 199–200
Thunderbirds, 139–40
*Tibicen*, 28–29, **29**, 99, 102
Tiger moth, 98, 150, 154
Tiger swallowtail, 94, 95, **Plate 33**
*Tillandsia usneoides*, 177
Timber rattler, 108, 114, **115**
*Tipularia discolor*, 55, 56, 83, 172
Torpor, 11, 21
Touch-me-not. *See* Jewelweed
Touchwood, 122
Towhee, 197–99
*Toxicodendron*, 48, 49, 72, 133–35
Trailing arbutus, 80, 173

Transylvania County, N.C., 6, 53, 147, 155, 199
Tree ears, 177
Treehoppers, 96
Tree rings, 19
*Trifolium repens*, 41, 42, **43**
Trillium, 4, 47–48, **Plate 5**
Tripe, 177
*Troglodytes*, 7, 199–200
Trout, 23–25, 27, 46, 110
Trout lily, 44, 46, 51, 72
*Trypoxylon politum*, 92–93, **92**
*Tsuga*, 167–68, **Plate 50**
Tufted titmouse, 141, 194, 196, 200
Tulip trees, 40, 56, 129, 151, 165, 166, **Plate 4**
Turkey brush, 175
Turkey vulture, 142
Turtles, 21, 117–19
Tympanum, 13, 14, 15, 28

*Umbilicaria*, 177
*Urocyon cinereoargenteus*, 189
*Ursus americanus. See* Black bears
Urushiol, 49
*Usnea*, 176–78, 185

Vampire bat, 146–47
Veery, 62, 134
Velvet ant, 88
*Veratrum viride*, 46
*Viburnum acerifolium*, 133
Viceroy, 93, **95**
Violets (*Viola*), 52
Vole, 29–31, **30**
*Vulpes vulpes*, 189

Wake-robin, 4, 47–48, **Plate 5**
Walnut, 135–37
Warblers, 4, 20, 61–63, 138, 141, **Plate 11**
Wasps, 78, 87–93, **92**, 101
Water moccasin, 114–15

Water snake, 114–15, **Plate 41**
Water witching, 127
Weather, 5–6, 155, 161–63
Weevil, 171
Whippoorwills, 140
White-breasted nuthatch, 197
White oak, 135–36, 138
White pine, 168
White squirrel, 137
Whitetail dragonfly, 101
White-tailed deer, 21, 131, 134, 170, 191, 192
White-throated sparrows, 199
Willows, 93
*Wilsonia citrina*, 62, **Plate 11**
Winter wren, 7, 200
Witch hazel, 121, 126–28, **127**, 156
Woodchuck. *See* Groundhog
Wood frog, **14**, 16, 111, 117, 201–2
Wood lily, 47
Wood sorrel, 12
Wood thrush, 5, 62, 134
Wood turtle, 117
Woolly alder aphids, 155–57
Woolly bear caterpillars, 154–55, **Plate 46**
Wrens, 7, 197, 199–200
*Wyeomyia*, 77

*Xanthorhiza simplicissima*, 75–76, **Plate 17**
Xylem, 18–19

Yellow-billed cuckoo, 65
Yellow fringed orchid, 55
Yellow jackets, 48, 78, 88–89, 187
Yellow poplar. *See* Tulip trees
Yellowroot, 75–76, **Plate 17**
Yucca, 81–82, **82**, 135
Yucca moth, 81–82, **82**

*Zonotrichia albicollis*, 199